SOUTHERN CULTURE ON THE FIZZ

SOUTHERN CULTURE ON THE FIZZ

AN EFFERVESCENT GUIDE TO
FERMENTED FOODS AND
BEVERAGES FROM
THE AMERICAN SOUTH

BRETT TAUBMAN

THE UNIVERSITY OF NORTH CAROLINA PRESS · CHAPEL HILL

This book was published with the assistance of the Blythe Family Fund of the University of North Carolina Press.

All photographs were taken by the author. All drawings courtesy of Lori Hill, 2023.

Designed by Lindsay Starr
Set in Sentinel, Din Next LT Pro, and Kiln
by Rebecca Evans

Manufactured in the United States of America

Cover art courtesy of Lori Hill, 2023.

Library of Congress Cataloging-in-Publication Data
Names: Taubman, Brett, author.
Title: Southern culture on the fizz : an effervescent guide to fermented foods and beverages from the American South / Brett Taubman.
Description: Chapel Hill : The University of North Carolina Press, [2025] | Includes bibliographical references and index.
Identifiers: LCCN 2024045137 | ISBN 9781469685410 (paperback) | ISBN 9781469685427 (epub) | ISBN 9781469685434 (pdf)
Subjects: LCSH: Fermentation—Southern States—History. | Fermentation—Amateurs' manuals. | Brewing—Amateurs' manuals. | Fermented foods—Southern States—History. | Fermented beverages—Southern States—History. | BISAC: COOKING / Methods / Canning & Preserving | COOKING / History | LCGFT: Recipes.
Classification: LCC TP371.44 .T38 2025 | DDC 641.4/630975—dc23/eng/20241211
LC record available at https://lccn.loc.gov/2024045137

For product safety concerns under the European Union's General Product Safety Regulation (EU GPSR), please contact gpsr@mare-nostrum.co.uk or write to the University of North Carolina Press and Mare Nostrum Group B.V., Mauritskade 21D, 1091 GC Amsterdam, The Netherlands.

I would like to dedicate this book to my family and the multitudes of students who have had to endure endless hours of my monologues and dad jokes. Now you, dear reader, get a small taste of what they have had to live through all these years.

CONTENTS

PART II. MEAT

PART III. DAIRY

ILLUSTRATIONS

Map

Tables

ACKNOWLEDGMENTS

I would like to express a sincere thank you to:

Lori Hill, for being a great friend in general, but especially for the amazing illustrations included in the book. Your beautiful artwork more than compensates for my terribly amateurish photography skills.

Julie Taubman, for being the best and most supportive wife I could ever hope for, and also for reading through multiple versions of the manuscript and providing me with your insightful observations.

Matthew Taubman, for generously dedicating your winter term project to the proofreading of the final manuscript.

Daniel Parker, for being an incredible lab and facilities manager. You keep the wheels turning on a daily basis. I have no doubt that you contributed to nearly every recipe in this book. Thanks as well for your unwavering willingness to try weird things on a daily basis with me. I couldn't do what I do without you.

All the students who not only contributed to this book but also endured my middle-aged-dad humor along the way. Special thanks to all the students who agreed to have their photos used for the book as well.

And, last, but certainly not least, Lucas Church and the rest of the UNC Press crew for helping me take this idea and turn it into something that is worthy of publishing.

SOUTHERN CULTURE ON THE FIZZ

A PRIMER ON FERMENTATION AND SOUTHERN CULINARY CULTURE

—

WHAT IS FERMENTATION?

You might think that as a professor of chemistry and fermentation sciences, this would be an easy question for me to answer. Alas, defining exactly what constitutes fermentation has plagued scientists for well over a century, roughly since the time microbes were first discovered to be responsible for fermentation.

As early as the seventeenth century, two eminent scientists were able to develop lenses and basic microscopes that made it possible to view microbes. Robert Hooke was an English polymath and the curator of experiments to the Royal Society of London who was credited with many scientific discoveries (some of which he may have stolen in his official role). Antonie van Leeuwenhoek was a Dutch microscopist who is also considered by many to be the "father of microbiology" and the fourth greatest Dutchman of all time (according to a public poll in 2004). Hooke is credited with coming up with the term "cells," to describe the honeycomb structure of cork seen under his microscope. More importantly, however, he was probably the first person ever to view individual microbial cells, as he described the hairy mold spots on cheese and several putrefying bodies in his seminal book *Micrographia* in 1665 (Donnelly 2014). Leeuwenhoek, by contrast, was also able to view microbes, which he dubbed *dierkins*, Dutch for "small animals" or "animalcules" as it was translated into English (Lane 2015), with his microscope, but that was not until 1674 (maybe that's why he's only the fourth greatest Dutchman?). However, another two centuries would pass before it would be confirmed that what these men were seeing were actually living organisms. Rather, in the case of a fermenting beer, the microbes observed were thought to be starch particles from wort, the sugar solution produced from malted cereal grains that is ultimately fermented into beer. Antoine Lavoisier, one of my favorite chemists who unfortunately lost his head (quite literally) during the French Revolution, was the person who figured out the chemical reaction of alcoholic fermentation, specifically the conversion of sugar to ethanol and carbon dioxide (Alba-Lois and Segal-Kischinevzky 2010). He knew—as did every brewer and winemaker for the past millennium or three—that yeast needed to be added to the sugar solution to make the ferment happen; but he had no idea that the yeast material added actually comprised living organisms. Instead, he conjectured that it was some sort of chemical catalyst that was necessary for the reaction to proceed and that was not consumed during the process. The yeast or "ferment" or "God is good," as it was referred to at the time, was recovered as a slurry at the bottom of the fermentation tank, or from the foam on top of an active fermentation, or even from paddles or other implements used to stir a prior ferment.

Over the next several decades, numerous advancements were made in our understanding of the fermentation process as well as to microscope technology, which allowed for better observation of microbes and their asexual reproduction, or budding, during fermentation. Despite these discoveries, chemists at the time (not surprisingly, as I'm well aware of the stubbornness of most chemists) continued to ridicule the scientists who had the audacity to believe that fermentation was controlled by tiny, single-celled organisms and not by a purely chemical process. One of the scientists who received a good bit of the chemists' ire was Louis Pasteur. He was able to demonstrate more conclusively the role of microbes in the fermentation process and that fermentation could not occur without microbes present. Additionally, while helping a distiller determine the reason for some bad fermentations, he noted that yeast cells were present in the fermentations that went well and produced alcohol, while there were far smaller cells present in the batches that did not produce alcohol. He determined that rather than alcohol, the primary compound that was produced in these batches was lactic acid. Thus, a new type of fermentation was confirmed, and instead of yeast, which was responsible for alcoholic fermentation, the much smaller bacterial cells were found to be responsible for this lactic acid fermentation.

Now we obviously have a much greater understanding of the fermentation process. The problem is that because there are so many microbes that can be involved, and so many different possible outcomes—depending on the microbes, the substrate being fermented, and the environmental conditions (temperature, pH, etc.)—we still don't have a clear definition of fermentation that everyone can agree upon. The name fermentation is derived from the Latin word *fervere*, meaning "to be hot, boil." If you've ever witnessed an active alcoholic fermentation, it is easy to see why this was the term selected for the process, as it can be a truly roiling, frothy solution of suds, not unlike a boil. Fermentation is generally considered to be an anaerobic process (in the absence of oxygen), and it was likely a metabolic adaptation by microbes to derive energy from carbohydrates in an environment lacking oxygen. However, there are a number of aerobic processes (in the presence of oxygen) that can be considered fermentation as well. So, we could almost settle on a definition of the fermentation process as any metabolism by microbes of biomolecules (starches, proteins, fats, etc.) in an anaerobic (or aerobic) environment; however, nonmicrobial processes that involve only the enzymes (proteins that act as catalysts of biochemical reactions) that would normally come from the microbes are also sometimes considered fermentation. Ugh!

Hopefully our graduates from the Fermentation Sciences program at Appalachian State University can help us to better define the process through a greater scientific understanding of the microbial world and our interactions with it. At the very least, they can spread the joys and wonders of fermentation throughout the southern United States.

WHY THE SOUTH?

The southern United States may not have the global financial powerhouse of a Manhattan or the celebrity glitz of a Los Angeles, but when it comes to culinary and musical traditions, I would put the South up against any other region in the country. As for the southern culinary traditions, I would even argue that it is the only truly unique and uniquely American region in the country. An amalgam of Indigenous, African, Creole, Caribbean, English, French, and Spanish influences, southern cuisine took some of the best of each of those cultures and transformed them into something that is distinctively southern. Don't believe me? If I mention barbecue, where do you think of? Sure, you may think brisket from Texas, or whole hog barbecue from North Carolina, but you think of the South. Yeah, yeah, Kansas City and their goopy, sweet ribs, blah, blah. How about shrimp and grits, a Low Country boil, country ham, blackened anything, jambalaya, fried chicken for crying out loud? I could do this all day. But what about the northern Midwest and their Scandinavian-influenced cuisine, or New York and the influence of Italian and Jewish immigrants, you say? The difference between those regions and the South is that the European influences didn't transform into something that is uniquely American. The American foods are, for the most part, very similar to their European counterparts. And don't get me started on Hotdish. Really, Minnesota, that's what you're going to hang your hat on?

It's not that the South had a monopoly on culinary creativity or a greater abundance of talented chefs (although nowadays that case could be made). Most of the foods that we recognize as southern today were born of necessity. Whether it was enslaved peoples who were provided with offal or the cuts of meat considered unsavory and who transformed them into what we now know as soul food, or southern farmers who relied on hogs that were easier and less expensive to raise than cattle and crops that were not grown elsewhere, like corn, collard greens, turnips, and sweet potatoes, these culinary practices were not by choice but were dictated by the exceptional circumstances and climate of the South. In the Northeast, by contrast, crops that were similar to what was grown in northern Europe were sown because the climates were more similar. As a result, the foods that were produced in that region did not differ substantially from their European influences. The South has a much warmer climate, however, so those traditional northern European crops could not be grown. For example, instead of wheat, southern farmers grew primarily corn as their main cereal grain. And now we have corn bread, a uniquely southern dish (Egerton 1993).

The hot climate and largely agricultural nature of the South also led to another southern culinary tradition—fermentation as a means of preserving foods. In the more northern climates, this wasn't as big of an issue because ice was more readily available. Even after the Industrial Revolution (which started in the South, of course, with the invention of the cotton gin), modern conveniences like refrigerators took longer to become commonplace in the South than in more urban centers in the North. As a result, southerners needed to preserve their harvests (both animal and vegetable) using the oldest form of food preservation. This is how we got traditionally southern foods like country ham, pickled fruits and vegetables like watermelon rind, okra, and beans, and of course bourbon (I mean something needed to be done with all that corn!).

THE BASIC SCIENCE

We are less ourselves than we are microbial. What the heck am I talking about, you ask? Well, we're carrying around more microbial cells than there are of our own cells. They're all over our bodies, inside and out, but the greatest population density is found in our guts. The microbial cells in our guts, which number in the trillions, are more numerous than the stars in our galaxy. And it's not like they're just along for the ride either. They are symbionts that do much of the dirty work our own cells are not capable of doing, the breadth and depth of which we're just beginning to figure out. For example, the microbes in our guts do much of the heavy lifting when it comes to metabolizing the food we eat. Without them, we wouldn't be able to extract the nutrients needed to stay alive. Similarly, when we ferment foods, we're using many of the same bacteria to break down the raw materials before they enter our bodies. It's almost like they're predigesting the foods for us to make them easier for us to further digest and access the nutrients. This is particularly helpful for some foods, like raw vegetables or dairy, that may be difficult to digest otherwise. So, how do the microbes metabolize the material? Well, you could suffer through multiple biochemistry courses like I had to and learn all of them, or you could read my succinct summary of two of the more important pathways we care about in fermentation—lactic acid fermentation and alcoholic fermentation.

For either pathway, we first must start with yet another metabolic pathway, glycolysis, which is the first step for much of metabolism in general. Why, you ask? Because this is the pathway that takes sugar and ultimately converts it into two pyruvate molecules and two ATP molecules. ATP, or adenosine triphosphate, is the molecule that is used to store the energy generated from metabolism in its chemical bonds. Pyruvate is the great metabolic intermediate, analogous to a train hub, that can allow metabolism to follow a number of different pathways depending on the need at the time. For us, that pathway is typically aerobic respiration. We live in an oxygen-rich environment, which is good, because we need the stuff to survive as we're aerobic organisms. This pathway allows us to extract the maximum amount of energy from a single molecule of sugar. That's all well and good, but our fermenting friends are generally anaerobic microbes. Most can survive in oxygen, but they can't necessarily use it for metabolism like we do. So, they have to use different metabolic pathways to generate energy. First of all, what role does oxygen play in metabolism? To understand that, we have to introduce another molecule, nicotinamide adenine dinucleotide (or NAD, which you may have heard of because it's one of the supplements du jour that supposedly makes you young again, balances your bank account, and takes out the garbage). Whereas pyruvate is the great metabolic intermediate and ATP is used to store energy, NAD is like the electron exchange market. NAD exists in two forms, oxidized NAD^+ (missing a negatively charged electron) and reduced NADH (with an extra electron in the form of the hydride ion, H^-), so it can receive electrons and then give them back, which is exactly the critical role it plays in metabolism. In glycolysis, two electrons are shifted to two NAD^+ molecules at one of the steps, converting the NAD^+ to NADH. The NADH molecules hold onto those electrons until they can drop them off with oxygen at the end of aerobic respiration. Thus, the NADH is converted back to NAD^+ that can then cycle back into glycolysis. But, without access to or the ability to use oxygen, what's a microbe to do? They need a means to re-oxidize the NADH to NAD^+, and they need to do something with the pyruvate that is produced during glycolysis. That's where lactic acid and alcoholic fermentation come into play.

Lactic acid and alcoholic fermentation provide alternative pathways to aerobic respiration when there is no oxygen present to act as an electron acceptor. They provide a means both to recycle the NAD^+ for glycolysis and to rid the cells of pyruvate. Lactic acid fermentation is a pretty straightforward process once the pyruvate from glycolysis has been produced, and

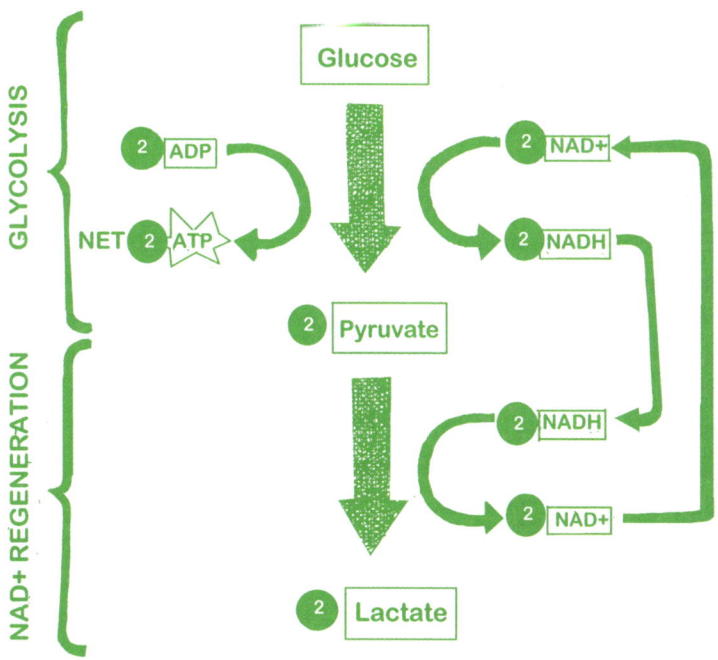

GLYCOLYSIS

NAD+ REGENERATION

Glucose

2 ADP

NET 2 ATP

2 NAD+

2 NADH

2 Pyruvate

2 NADH

2 NAD+

2 Lactate

Glycolysis followed by lactic acid fermentation.

it is commonly employed by lactic acid bacteria as well as by our muscle cells when they lack oxygen after rigorous exercise. To regenerate the NAD$^+$ necessary for glycolysis, the electron from the NADH is transferred to the pyruvate in the form of the hydride ion, which converts the pyruvate to lactate, which is then excreted from the cell into solution. The NAD$^+$ is then available again to be fed back into glycolysis and the cell has rid itself of the pyruvate.

Alcoholic (or alcohol) fermentation is only slightly more complicated than lactic acid fermentation and is used by anaerobic yeast in the absence of oxygen. This time, the first step involves decarboxylating (a fancy word that means removing a carbon dioxide from a molecule) the pyruvate. This gives acetaldehyde, a two-carbon molecule (since one carbon has been removed from pyruvate with the carbon dioxide). Now, the NADH has another willing receptor for its electron, the acetaldehyde. In doing so, the acetaldehyde is reduced to ethanol (alcohol) and the NAD$^+$ is restored for cycling back into glycolysis. Similar to lactic acid fermentation, the ethanol is excreted from the cell, as it would otherwise be toxic to the yeast, and everyone is happy (especially those who get to consume the alcohol). Acetaldehyde, by the way, is the chemical compound that leads to hang-overs when it builds up in our systems. When we consume alcohol, it is metabolized by first breaking it down to acetaldehyde (the reverse of how it was made by the yeast). These reactions, both the production of ethanol by yeast and the breakdown of it in our bodies, would not occur without the assistance of the alcohol dehydrogenase enzyme complex. The more alcohol you consume on a regular basis, the more this enzyme is upregulated to help metabolize the alcohol, since it is toxic and needs to be removed from our system (that is, up to the point of severe alcoholism and liver failure when the body can no longer efficiently produce the enzyme). Some ethnicities, such as people of East Asian descent, have semi- or nonfunctional alcohol dehydrogenase enzymes (Wall and Ehlers 1995). People with this condition often flush around the neck and face when consuming alcohol and may suffer from raging hangovers, both symptoms resulting from the buildup of acetaldehyde in their systems. So, putting on my dad hat here, make sure that if you do consume alcohol, do so in moderation and know the limits of your body.

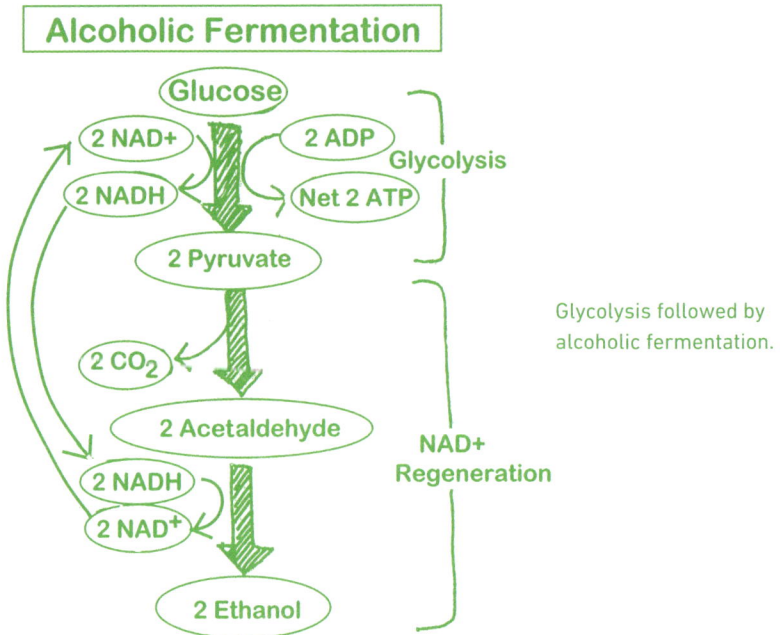

Glycolysis followed by alcoholic fermentation.

BACTERIA

Before delving deeply into a detailed discussion of the single-celled organisms responsible for fermentation, it will probably be helpful to take a step back and differentiate the two types of living cells possible, prokaryotic and eukaryotic cells. Prokaryotes are the OGs. They probably evolved around 3.8 billion years ago and were the first living organisms to grace our planet. Eukaryotes came later, likely about 2.7 billion years ago through the phenomenon known as symbiogenesis and described below. Prokaryotes do not have a nucleus, just free-floating DNA, unlike eukaryotes that have a defined nucleus as well as other membrane-bound organelles, something that prokaryotes are also lacking. Prokaryotes are always single-celled organisms and include bacteria and archaea. They reproduce only asexually through binary fission, a fancy way of saying the cell replicates its DNA and splits in two, producing two genetically identical daughter cells. Eukaryotes can be single-celled organisms, like yeast, or multicelled, like humans and cats and trees; and eukaryotes can reproduce either asexually through cell division (mitosis and meiosis, you know, those processes you learned about in high school biology but have no reason to ever remember again) or sexually (thankfully, as I prefer this method over budding, personally). Finally, prokaryotes are much smaller than eukaryotes, like 10 to 100 times smaller in some cases, which doesn't matter except when you're trying to teach people how to use a microscope to observe bacteria and yeast or when you're trying to filter your fermented beverage and need to select the correct pore size. And now let's talk about the largest group of microorganisms, the bacteria.

There are thousands of species of bacteria, the majority of which are harmless, although some can be pathogenic. Most of the bacteria we're concerned with are gram-positive bacteria, but some fermentations do involve gram-negative bacteria as well, so those will be discussed briefly, too. You may be wondering right now what the heck I mean by gram-positive and gram-negative. Gram staining is typically the first step used in generally identifying bacteria. It's a simple staining procedure that stains cells either purple (gram-positive cells) or pink (gram-negative cells) based on the composition of their cell walls. Identifying bacteria as gram-positive or gram-negative allows a scientist or medical professional to narrow down the possible species in question.

9

Lactic acid bacteria (LAB), the bacteria typically involved in fermentation, are gram-positive. They are either rod-shaped (bacilli) or spherical (cocci), and most species are aerotolerant anaerobes, meaning they can tolerate a little oxygen; but don't expect them to use it to metabolize sugars into energy (otherwise known as aerobic respiration in case you care). Some species may, however, be facultative anaerobes, which actually can use oxygen to generate energy from sugars. LAB are acid tolerant, which is not surprising considering the main product of their fermentation is (shocker) lactic acid. Many species are homofermentative, meaning that lactic acid is the main product of fermentation, whereas others are heterofermentative and produce acetic acid, ethanol, and carbon dioxide as well as lactic acid during fermentation. Some species can go either way depending on the environmental conditions, such as pH, temperature, and available carbohydrates. Many LAB are also halophilic or at least halotolerant, indicating that they either need sodium chloride (table salt) to survive and/or can handle pretty large concentrations of the stuff, like upward of several percent. This, together with the fact that they produce lactic acid (and sometimes acetic acid) and are generally recognized as safe (GRAS), makes them the ideal food fermenters. When foods are heavily salted, this naturally selects for the halotolerant LAB, allowing them to get a leg up on metabolizing the food. As they do this, they use up the available nutrients, thereby further limiting the ability of other microbes to proliferate. In addition, they produce acids, which themselves are antimicrobial and drop the pH to a level at which most other bacteria cannot survive. Some also produce bacteriocins, natural antibiotics (similar to penicillin, although this is produced by a mold) that inhibit other bacteria. For all these reasons, LAB are the most commonly used bacteria in fermented foods. They are found in yogurts, cheeses, sourdoughs, pickles, sauerkraut, and many, many more items. LAB are also some of the gut microbes we want to encourage and the bacteria found in many probiotic formulations.

There are over a dozen different LAB genera. The ones that we're mainly concerned with, though, are *Lactobacillus*, *Lactococcus*, *Leuconostoc*, and *Oenococcus*, with a few more bit players. There are way too many species to cover all of them, but I'll hit on some of the more important ones for food production:

- *Lactobacillus plantarum, Lactobacillus brevis, Pediococcus pentosaceus, Pediococcus cerevisiae,* and *Leuconostoc meserenteroides*: These are some common species that are often found naturally on vegetable matter or that are used as inoculants for vegetable ferments such as pickles, sauerkraut, and kimchi.

- *Lactobacillus plantarum*: These are isolated from naturally fermented meat products and are commonly included in starter cultures for meat fermentations.

- *Lactobacillus sakei* and *Lactobacillus curvatus*: These are two other species used in meat fermentations.

- *Pediococcus acidilactici*: These bacteria are not found naturally in meat ferments but are useful for higher temperature, fast meat fermentations.

- *Lactococcus lactis* ssp. *cremoris* and *lactis* and *Leuconostoc meserenteroides* ssp. *cremoris*: These species are used in buttermilk production. *Lactobacillus lactis* ssp. *cremoris* and *lactis* are also common mesophilic (moderate temperature range) species used in cheese production.

- *Streptococcus thermophilus* and *Lactobacillus delbrueckii* ssp. *bulgaricus*: These are the essential flora of yogurt and are common thermophilic species used for cheese production as well.

- *Oenococcus oeni*: These are the LAB involved in wine malolactic fermentation, which converts the harsh malic acid to the softer lactic acid, a flavor profile preferred in many red wines and some white wines.

LAB are often found in beer as well, whether desired or not. Many traditional styles as well as some more modern ones use mixed-culture yeast and bacterial fermentations to achieve more complex and acidic beers. The LAB found in beer and breweries are primarily *Lactobacillus* and *Pediococcus* species, including my favorite, *Pediococcus damnosus*, because when you find it in your beer, your response is to look to the heavens and shout "damnosus!"

Acetic acid bacteria are another popular group of bacteria used for specific fermentations. These are gram-negative bacteria from the family *Acetobacteraceae*, which includes the two acetic acid bacteria genera, *Acetobacter* and *Gluconobacter*. These bacteria are rod-shaped and need oxygen to grow and metabolize (obligate aerobes for the three people reading this who actually care). Not surprisingly, acetic acid bacteria metabolize ethanol to produce acetic acid. This is how vinegars are made, through the production of wine or other alcoholic beverages, followed by the exposure to oxygen and acetic acid bacteria. Also not surprising, since they produce acid, is the fact that they are acid tolerant, making them perfectly suited to the fermentation life. They are very common in nature: pretty much anywhere a sugary solution is found, or on overripe, bruised fruit, even on fruit flies, which is why we have a strict fruit fly eradication program in our facility come summertime (that may or may not involve me with a spray bottle of ethanol and a torch, but which I am not allowed to say definitively for legal reasons).

When vinegars are being fermented, they often grow a "mother" on top of the fermenting solution. This mother is similar to a kombucha SCOBY, which is an acronym for Symbiotic Culture of Bacteria and Yeast. The SCOBY, or mother in this case, is produced specifically by the acetic acid bacteria. It is largely just cellulose and made from the extracellular polysaccharides that the bacteria excrete, which is a fancy way of saying that the bacteria poop out cellulose (or would if they had butts, which they don't, so instead they "excrete"). This cellulose raft floats on top of the solution, which provides an environment with access to the oxygen in the headspace above the solution so that the aerobic acetic acid bacteria can ferment the solution below while still able to access oxygen. The SCOBY, or mother, is perfectly food-safe although not very digestible. If the ferment is allowed to go for a while, "daughters" will be produced as well, which are just new cellulose rafts that are produced by the bacteria. The daughters form on top of the mother, which then typically sinks to the bottom of the solution. After a while, there may be multiple layers of discs that form and sink, with the newest daughter floating on top of the solution. The mothers, or SCOBYs, can be used to inoculate subsequent batches, but they are generally filtered from the liquid before packaging and selling the vinegar or kombucha. Occasionally, however, you will find vinegars or kombuchas that have live cultures with a mother, or SCOBY, present in the bottle. Since you can't really eat them (well, technically you can eat them, it's the digesting part that is the problem), people have come up with all kinds of fun uses for

these cellulosic discs. After they are dried out, they can be made into fris-
bees or even vegan "leather" clothes and furniture. I'm personally thinking
of starting a line of vegan, biodegradable yarmulkes.

Because many wines are aged in oak barrels for extended periods of
time, it is virtually impossible to keep them anoxic. Nor would you want
to, as that oxidation is part of the process. However, as a result of having
your wine exposed to oxygen for so long, it is critical to guard against infections by acetic acid bacteria. This is done by using proper concentrations
of sulfur dioxide as a preservative, which prevents the growth of acetic acid
bacteria. There are actual legal regulations for the amount of volatile acidity that can be found in wine. The volatile acid that is the real concern is
acetic acid. If you have too much of that, you no longer have wine, you have
vinegar.

YEAST

Yeasts are single-celled, non-photosynthetic eukaryotic fungi, part of the
same kingdom that includes molds and mushrooms. They evolved hundreds of millions of years ago and are the result of an earlier event referred
to as symbiogenesis, which really has nothing to do with the book, but is one
of the coolest evolutionary adaptations I know of, so let's do a little sidebar
action on it.

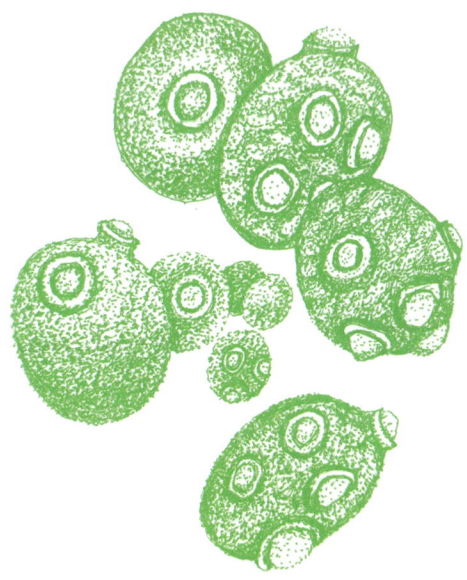

Symbiogenesis

—

Symbiogenesis likely occurred around 1.5 billion years after the emergence of prokaryotic cells. At some point, one prokaryotic organism oozed its way through the cell membrane of a larger one, either as prey or as a parasitic organism. Somehow, though, rather than consuming each other, they figured out how to live together in a symbiotic relationship. As a result, an aerobic protobacterium that could use oxygen to make energy ended up becoming mitochondria, which helped the larger organism thrive in an increasingly oxygen-rich environment. Similarly, cyanobacterium, which are capable of photosynthesis, became another endo-symbiont and in green plants evolved into chloroplasts, the organelles that perform photosynthesis in modern plant cells (Sagan 1967). I feel like there's a metaphor in here for modern society. How is it that single-celled organisms are more capable of sorting out their differences? Well, maybe because they've been around for over a billion years longer than we have. As they say, time heals all wounds. •

Over 1,500 distinct species of yeast have been identified. This is such a large and diverse number of species that they actually belong to two different phyla, the Ascomycota (which includes the Saccharomycotina or true yeasts) and the Basidiomycota. The yeasts we are mainly concerned with here belong to the genera *Saccharomyces*, although some wild type yeasts from other genera that play important roles in fermentation processes will be discussed briefly below. Yeasts are chemoorganotrophs, a fancy word meaning they consume organic compounds (like sugar) to extract energy and can be either obligate aerobes that require oxygen for cellular respiration or facultative anaerobes, which means they prefer an anaerobic environment but are capable of aerobic respiration in the presence of oxygen. There are no known yeast species that are obligate anaerobes, which can survive and grow only in environments devoid of oxygen. We're not worried about the obligate aerobes because those aren't the species that produce ethanol. The facultative anaerobes are the yeasts that developed the alcoholic fermentation pathway to extract some energy from sugar while also ridding the cell of toxic metabolic intermediate species. A win-win for the yeast cell. And win-win-win for us, since ethanol is the end product of alcoholic fermentation.

Most yeasts prefer slightly acidic environments and are found at virtually every temperature range, from below freezing to well above 40°C. Given these simple ecological requirements, it's no surprise that yeasts are found in a number of different environments. However, *Saccharomyces* species are generally niche organisms in the environment, largely residing on sugar-rich materials like the skins of fruits. Considering that they are the most widely used microbe in the world, this is quite the step up for these little critters. I always chuckle when people talk about the domestication of yeast, like all the other animals we have domesticated. I would argue it's the other way around. Yeasts have done a pretty good job of domesticating us. They have gone from a niche organism to captivating humans to the point that we gave up our hunter-gatherer ways for an agricultural society with entire facilities dedicated to the feeding and proliferation of yeasts for the production of massive quantities of deliciously addictive beverages. Even if nobody else sees it, I'm on to you, yeast. Is this like a microbial *Matrix* situation? If it is, I think I'm OK with it.

A Little Brewing Yeast Genetics

Most wild yeast species are haploid, meaning they only have one set of chromosomes. However, most brewing yeasts are polyploid, with multiple copies of their DNA material. Humans have been brewing beer from cereal grains for thousands of years now. And quality control isn't something that was recently invented. Sure, we have a better understanding of the science, especially microbiology, and there are more stringent control mechanisms in place, but there has been a lot of pressure to make good beer for a long time. Nowadays, if you make bad beer, you'll probably go out of business (unless you make hazy IPAs—then you can pretty much make any old swill and get away with it). Back in the day, bad beer may have led to a beheading, so yeah, there was a lot of pressure to make good beer.

A large part of making good beer is being consistent. If you make a good batch, how can you replicate it and make it again and again. Even before we had any clue what yeasts were, we understood how to transfer the part of the previous ferment (which happened to be the yeast) to the current one to initiate fermentation. Of course, only the batches that turned out well would be used in this manner. The brewers were unwittingly selecting for the most successful yeast for brewing. Cells mutate all the time. When you're a multicellular organism, mutations in individual cells usually aren't even noticed. It's only when lots of cells begin to mutate, like with cancer, that we begin to notice. With single-celled organisms, single mutations have much larger impacts. At some point, some cells mutated to carry multiple copies of their DNA. The ones that suddenly had multiple copies of the MAL gene, the gene ultimately responsible for the uptake and metabolism of maltose and maltotriose, the two most common sugars found in beer mashes made from cereal grains, were capable of consistently metabolizing those sugars and making good beer. If a yeast had only one copy of the MAL gene, like wild type yeasts, it would not be able to make consistently good beer. This is because, if a mutation were to occur in or around that portion of the DNA that rendered the MAL gene dysfunctional, then that yeast wouldn't be able to properly metabolize the sugars in the fermentation solution. By contrast, if a previously mutated yeast has multiple copies of the MAL gene, even if a further mutation were to occur on one or a couple of the chromosomes that would render the MAL gene inoperable, there would still be more copies of the MAL gene on the other chromosomes to pick up the slack. By using the yeast that most consistently fermented good beer over many generations, brewers had selected those strains that had multiple copies of their chromosomes with multiple sets of MAL genes. •

Two main species are used for beer production, *Saccharomyces cerevisiae* and *Saccharomyces pastorianus*. These yeast species and their lascivious love lives are discussed in greater detail in chapter 17, which includes an excerpt from my upcoming romance novel, *Beauty and the Yeast*. Wine production mainly uses *S. cerevisiae*, which is inoculated into the grape must. However, most grape skins harbor several different species of wild yeasts, including *S. cerevisiae* as well as yeasts from the genera *Kloeckera* and *Candida*. Traditionally, these would be the only yeasts used for inoculating the grape must (those extant on the grape skins). Even now, when most winemakers pitch a controlled amount of purchased, dried yeast into their grape must, the wild yeasts that are present on the skins already will also play at least a small role in the fermentation. Additionally, *Saccharomyces bayanus* is an alcohol tolerant yeast species that is commonly used to inoculate certain high Brix (sugar) grape musts or in the production of fortified wines. *S. cerevisiae* is also the most common species used for cider production and in the mixed-culture fermentations of kombucha. A wild yeast that is also associated with both cider and kombucha ferments as well as with beer and wine production (although not always intentionally) is *Brettanomyces*.

Brettanomyces is a love it or hate it type of yeast. This has nothing to do with its brash personality or immature sense of humor (although it is known for both, oh wait, no, that's me) but rather with the flavors it produces during fermentation. These flavors have attractive descriptors such as barnyard, horse blanket, leather, smoky, band-aid, spicy, the list goes on, but I think you get the point. I, personally, am a huge fan. And, no, it's not because the yeast is my namesake. It's not by the way, although I did consider lengthening my name to Brettanomyces to spice things up a bit. I just really appreciate the funky flavors that this yeast produces, even though many people, understandably, have a less-than-favorable opinion of the same flavors. In most wineries, simply uttering the name *Brettanomyces* is enough to cause winemakers to strip off their clothes and run screaming for the nearest safety shower. It has historically been considered a fault if *Brettanomyces* and its characteristic flavors end up in wine, although it is often associated with many traditional European styles. Nowadays, some wineries are embracing the funk and even inoculating their wines with *Brettanomyces*. Many breweries have been doing this for quite some time; however, many brewers fear the funk as well, and for good reason.

Brettanomyces is a non–spore forming yeast that can metabolize sugars under both aerobic and anaerobic conditions. It will produce large amounts of acetic acid (the acid that makes vinegar) in glucose rich solutions under aerobic conditions. This yeast metabolizes sugars much more slowly than *Saccharomyces*, so it may not be evident in the early stages of fermentation. However, *Brettanomyces* can ferment a much wider range of sugars and starches than *Saccharomyces*, and some species can even metabolize ethanol; so if there is some *Brett* in your beer or wine, it will eventually show up and start to impact the flavor profile. *Brettanomyces* can form protective biofilms and live deep inside of wood barrels (some species can metabolize the cellulose and other wood sugars in the wood), which makes it extremely hard to eradicate once it has established itself in a production facility. In the natural environment, it lives on fruit skins and may be transported by small insects like fruit flies. In traditional British ciders, *Brett* character would often be present because of the yeast living on the skins of the apples. Fortunately, its boat-shaped or elliptical morphology makes it easy to identify and distinguish from *Saccharomyces* yeast using light microscopy, and it can be controlled with sulfur additions. That said, more wineries, breweries, and cideries are beginning to embrace the funk in small doses. To that I say, long live *Brett* (the yeast, not me, but well, me too)!

KOJI

Koji technically refers to the mold, *Aspergillus*, and the substrate it has grown on. I'll switch back and forth between koji and *Aspergillus* depending on the context, but for the most part I will use koji, mostly because it's easier to type and I don't have to italicize it. *Aspergillus* itself is a filamentous fungus (a fancy euphemism for mold) that has been used for millennia in East Asia for food and beverage production. Right now, you may be thinking, mold for food production? But I thought moldy food should get thrown out. And, you're right: if you don't know the mold growing on food, you should definitely throw it out. Molds can produce mycotoxins, which can cause bodily harm or even death in some extreme cases (Egbuta, Mwanza, and Babalola 2017). There are a few primary molds associated with foods, *Aspergillus*, *Penicillium*, and *Rhizopus*. When your oranges or bread get moldy, it's probably *Penicillium* (although it could be *Rhizopus*). And *Penicillium* is commonly used in cheese and fermented sausage production as well as in the production of the antibiotic penicillin. If your strawberries, tomatoes, or sweet potatoes grow mold, chances are it's *Rhizopus*, which is also the mold used for tempeh production. However, *Aspergillus* is likely the most commonly used mold worldwide. Some of the products made with koji include miso, shoyu, sake, shochu, and rice vinegars, to name a few. Regardless of their ubiquity, molds have good species and strains that are used in food and beverage production and bad ones that should be avoided. It's virtually impossible to distinguish a good mold from a bad one with the naked eye, however. That's why it drives me crazy when I read advice regarding fermentations or other food products stating that if the mold is white (or maybe pink or green), it should be OK. Just scrape it off and pretend that it never happened. Nothing to see here. But if it's black, that means it's bad. However, if you didn't intend for the mold to be there, then you really have no idea what it is. And mold color means nothing with respect to its toxicity. The same mold can start off one color and change over time. Or the organism may produce different colors of molds depending on the environmental conditions, such as pH and humidity. Many strains may be fine to use under certain environmental conditions but can also produce harmful mycotoxins under different conditions. *A. niger*, for example, is used in the food and pharmaceutical industries, but it is also the black mold that invades homes and can cause illness. To be completely safe, you should probably discard the food item. However, there are certain instances where it may be fine and you can probably just scrape it off, treat it with some vinegar and salt (more on this in the "Safety First" section of this chapter), and move on.

For example, there are many times when we're fermenting/aging different meats in one of our fermentation cabinets. Some may be treated with mold to protect the casing, and some may be smoked (also to protect the casing). Occasionally, the untreated, smoked meat product may start to show some mold spots. In that case, since we're pretty certain the mold is derived from one of the strains we intentionally inoculated on another product, we can assume it's not harmful, especially if we remove it as quickly as possible. In general, the substrate plays a big role in the potential-harmfulness factor of molds. If the mold is growing on a sausage casing for example, especially if that casing is not going to be consumed, the mold can probably be safely removed without any deleterious consequences, if it is removed early. However, if you're making a crock of kraut and you get a nice fuzzy layer of mold on top, throw it out immediately. There is way too much liquid in solution during a vegetable fermentation. Any mycotoxins produced will dissolve right into that solution and cannot be removed after the fact. A meat ferment, by contrast, has a lot less liquid by which to distribute any mycotoxins, especially if the mold is growing on a casing. And if that casing is going to be removed anyway, it should be fine. So, context matters, but if you're unsure about what to do based on this meager advice, err on the side of caution and throw it out. It sucks when you have to throw away food items, but if it's the difference between losing a few bucks and lifelong liver damage, take my money!

A. oryzae is the most common species of *Aspergillus* mold used in food and beverage production, but *A. niger*, *A. kawachii*, and *A. luchuensis* are also frequently used. In general, koji is used for the enzymes that it produces and excretes to degrade the substrate. It is typically used in conjunction with other microbes, either bacteria or yeast. In sake production, for example, polished rice is treated with *Aspergillus*. The amylolytic, or starch degrading, enzymes produced by the *Aspergillus* degrade the starch molecules in the rice to simple sugars that can be fermented to ethanol by the yeast that is added later in the process. In fact, out of all eukaryotic organisms, *Aspergillus* produces some of the highest levels of alpha-amylase, one of the primary starch-degrading enzymes secreted via an organelle called the Spitzenkörper (which is not at all relevant to this discussion, but it's my favorite name for an organelle, so there you have it) (Kitamoto 2015). Enzyme production, however, is dependent on strain type and environmental factors such as temperature and humidity.

Jokichi Takamine, a Japanese-born chemist who emigrated to the United States in the late nineteenth century, was the first to isolate and purify alpha-amylase, which he dubbed takadiastase, from koji. His mother was from a family of sake brewers, so I'm sure he was very familiar with koji. After moving to the United States, he sold the production rights to his takadiastase to a pharmaceutical company and made a rather large fortune. He was also the first to isolate and purify the hormone adrenaline. After emigrating to the United States, he made it his life's mission to serve as a goodwill ambassador between Japan and his new homeland. With that in mind and all his accumulated wealth, he donated, along with the mayor of Tokyo, the cherry blossom trees in the West Potomac Park in Washington, DC. This is a story of koji, though. Takamine was probably the first person to apply the enzymatic properties of *Aspergillus* for commercial uses after the isolation of alpha-amylase (Schoene, Fulmer, and Underkofler 1940). He developed a commercial process for using koji to produce whiskey. His research showed that koji used on raw grains in a whiskey mash resulted in an efficiency comparable to malted grains (Takamine 1914). However, somebody must not have been happy with his foray into whiskey production, as his lab mysteriously burned down one evening, which put an end to that line of research. Fortunately, his advancements and innovation are not forgotten, even in the whiskey industry. If you go to Rabbit Hole Distillery in Louisville, Kentucky, one of the newest distilleries in Louisville, there is a large mural honoring the forgotten distilling heroes of the past. Jokichi Takamine's name is front and center, as you can see in the photo of the ever-fashionable author smiling next to the mural like a school kid who has just met one of his heroes.

GEOTRICHUM, PENICILLIUM, AND RHIZOPUS, THAT IS, THE OTHER MOLDS

Geotrichum is another common mold genus that is found seemingly everywhere in the environment, including in water, air, soil, sewage, plants, and dairy products as well as in us. That's right, it's often isolated from the skin, lungs, and gastrointestinal tract. Typically, it's not pathogenic, but in rare cases it can lead to a potentially harmful infection known as geotrichosis, a condition often associated with those who are immunocompromised (Keene et al. 2019). It is a saprobic (decomposer) organism that can occasionally become parasitic (feeding off a living host, as when it's pathogenic) with over 100 known species. From an evolutionary perspective, *Geotrichum* is a particularly interesting organism. All true yeasts, which are single-celled fungi, evolved from molds, or filamentous fungi, which are technically multicellular organisms. When a yeast cell reproduces asexually, it buds, and then sheds the daughter cell to form another single cell.

The fashion icon/author standing next to a mural of forgotten distilling heroes of the past, including Jokichi Takamine, in Rabbit Hole Distillery in Louisville, Kentucky. Yes, those are pineapples on my shorts.

Molds, by contrast, are made of a network of filamentous threads (hyphae), that reproduce by growing and dividing from the end of the thread, but not necessarily separating, thus extending the thread. Eventually, these threads become quite long and branched, intertwining and forming a vast network referred to as mycelium. This is what gives mold its fuzzy appearance. Well, *Geotrichum* can't really make up its mind. When *Geotrichum candidum*, the common cheese mold, grows on its cheesy substrate, it certainly looks like any other white, fuzzy mold. However, when viewed under a microscope, it often appears as single-celled organisms, just like yeast. Why the evolutionary conundrum? Well, it seems that *Geotrichum* shares many genes with both modern yeasts and their fuzzy progenitors. One important genetic clue to its appearance is the fact that *Geotrichum* has many more genes that code for the production of chitin, the polymer that is a major component of yeast cell walls (as well as a major component of the exo-skeletons of arthropods). Chitin is critical in producing the mycelium associated with molds, which is why *Geotrichum*, with an excess of chitin, often produces the filaments that make it appear to be more mold than yeast. Similarly, *Geotrichum* produces a ton of lipases, the lipid (fat)-degrading enzymes that lead to the characteristic flavors in mold-ripened cheeses, which are also found in other molds but not necessarily in yeast genomes (Wolfe 2023). Excreting enzymes is how molds feed. We produce enzymes as well, but we do so in our gastrointestinal tract, and these enzymes help to break down foods into individual molecules that can then be absorbed and transported where needed in our bodies. Molds, by contrast, excrete enzymes from their mycelium into their environment. The enzymes do the same thing that ours do (break down food into individual molecules), but they do so in the environment surrounding the mycelium. Then the molecules are taken up by the mycelium and used to feed the growing mold. This trait makes certain molds particularly useful in food production. In this case, it makes *Geotrichum candidum* a great cheese mold. The mold excretes lipases into the cheese, breaking down the fats into short-chain fatty acids, which have the funky, mushroomy, goaty flavors often associated with mold-ripened cheeses.

Penicillium is a mold that most people are at least familiar with. Or at least familiar with one of its important contributions to medicine, penicillin, the antibiotic produced by some members of the genus that kills or prevents the growth of many kinds of bacteria. It is a very common mold, with over 300 species, that is used industrially to produce pharmaceutical compounds and is found throughout the environment, including on foods. This is the application we care about, namely its use in cheese and sausage production. *Penicillium candidum* and *Penicillium camemberti* are commonly used with mold-ripened cheeses such as Camembert and Brie. They produce the white, fuzzy mold that makes the brine on these cheeses. The proteolytic (protein degrading) and lipolytic (fat degrading) enzymes excreted by the mold help to ripen these classic cheeses and give them their distinctive flavors. *Penicillium roqueforti* is the mold used to produce the characteristic blue veining and flavor in blue cheeses. *Penicillium nalgiovense* is used to make some soft, mold ripened cheeses but is more commonly recognized as the white mold used to coat the casings in sausage production. Not only does the mold protect the casings and prevent other molds from growing but also the enzymes that are excreted help to break down the macromolecules in the meat and provide flavor to the finished product.

Additionally, the mold helps to dry the meat, making it suitable for human consumption, as the mycelium sucks up the moisture. At the same time, though, it also helps to equilibrate the moisture in the meat, preventing the formation of a dry rim to the sausage. Finally, *Penicillium nalgiovense* also degrades the lactic acid produced by bacteria, thereby increasing the pH and preventing an overly sour flavor from developing in the sausage.

Rhizopus is another saprobic fungus with somewhere around 10 distinct species. It is used industrially as well as in food production, but it can also be pathogenic to both plants and animals. It is known colloquially as the common bread mold, since, and you may not have seen this coming, it grows on bread (as well as on lots of other foods). Two species in particular, though, have been used in Asia for centuries to produce food as well as alcoholic beverages. *Rhizopus oligosporus* is the species more commonly used to make tempeh from soybeans, although it may actually perform better with tempeh produced from beans other than soy. Tempeh, the fermented

Top: The beautiful white *Penicillium nalgiovense* coating salami.

———

Left: A ripening camembert covered in *Penicillium* and *Geotrichum* molds.

23

soy product originating from Indonesia, is relatively easy to produce, inexpensive, and fairly flavor neutral, making it adaptable to many different recipe inclusions. And it's really friggin' good for you. Soy, in general, has many compounds that have been shown to be healthful, including having anticancer properties, but ferment the soy and the healthfulness only increases. Fermentation increases the number of bioactive compounds and their ability to prevent or fight cancer among other benefits (Nurkolis et al. 2022). A study in 2019 showed that the bioactive peptides in tempeh had antihypertensive, antidiabetic, antioxidative, and antitumor properties (Tamam et al. 2019). So, now that you've cured yourself of any ailment, it's time to wash down that tempeh with some alcoholic beverages made from *Rhizopus oryzae*. Whereas this species can become an opportunistic pathogen in humans, causing mucormycosis (or black fungus), although primarily in those who are immunocompromised, it is considered GRAS by the FDA and has plenty of industrial applications. These stem mainly from the fact that it produces not only a lot of starch-degrading enzymes but also organic acids, fruity esters, and ethanol, which make it a great option for producing traditional rice wines in Indonesia, China, and Japan. This species is also great for making soy-based tempeh.

SAFETY FIRST

In this section, I will discuss the importance of sanitation as well as recommendations for how to properly sanitize your workspace and equipment. I will cover the proper use and concentrations of salt to select for good microbes, why pH is so critical to the process and how to monitor it, sensory clues that indicate whether the fermentation is proceeding properly or not, and other issues to monitor to ensure safe and healthy ferments.

Fermentation is a process that has been employed by humanity for millennia to preserve food and make it more healthful and delicious. That said, you are relying on microbes to do the work, and you are often selecting for beneficial bacteria over pathogenic bacteria in the environment; so, following some basic safety guidelines is critical to the success of your ferment as well as to your health. There are too many so-called fermentation experts out there with no scientific backgrounds who eschew basic safety protocols. To put it bluntly, their laissez-faire advice regarding safety (or lack thereof) is going to seriously harm or possibly kill someone. So, now that you're suitably terrified, I am here to remind you that home fermentation is easy and perfectly safe—as long as you follow basic safety protocols.

When it comes to fermentation safety, there are three basic rules to follow that are a great start to ensuring a safe ferment: (1) clean, (2) sanitize, and (3) sanitize. By cleaning, I mean removing all dirt and debris from any of the materials that will come into contact with your ferment. This could be accomplished with soap and water, but some longer-term ferments may leave behind some pretty intractable residues. For those, an alkaline, non-chlorine oxidizing cleaner, such as Powdered Brewery Wash (PBW™) by Five Star, is very effective. Also effective (and cost-effective to boot) is washing soda, which can generally be found in most larger box stores. Washing soda is just sodium carbonate, a more alkaline compound than its cousin, sodium bicarbonate, which you may already know is baking soda. Before you develop tendonitis trying to scrub off that debris, give the materials a nice soak in a washing soda solution. That debris will come right off. Now that your materials are cleaned, they must be sanitized. This is the process whereby most of the microbes on your materials are killed. This is not full sterilization, in which all the microbes are killed. Sanitization decreases them to a safe level, meaning that there are so few, they won't have a chance to compete with the microbes involved in the fermentation and cannot reach pathogenic levels. For sanitizing, I prefer food-grade, no-rinse, chlorine-free sanitizers. Star San, also by Five Star, is a high foaming, no-rinse, acid-based sanitizer. The foam helps to penetrate those hard-to-reach nooks and crannies. It only takes a few minutes of contact time to work and then just needs to be drained. No rinsing necessary. I stay away from chlorine bleach. Not only does it potentially lead to some nasty compounds and off-flavors if there is any residue remaining, but you have to rinse after application. In my opinion, this negates the sanitization process. You just spent that effort trying to remove the microbes, so why are you rinsing with water from a faucet, both of which may harbor microbes. Sure, if you're on a municipal supply, there's not much likelihood that your water will have microbes, but these days, who knows. And I can guarantee that your faucet dispenser is carrying a good number of microbes (which is why you should clean your faucets regularly!). Why take the risk? Plus, chlorine is corrosive to several materials that you may be using, and it will decrease the lifespan of those materials. I also avoid iodophor, which is a no-rinse solution of bound iodine. The iodine, which is released in solution, is a very effective sanitizer. However, unless you want all your materials to stain purple and to have a metallic off-flavor introduced to your ferments, I would avoid it.

Adding the proper amount of salt to your raw materials is one of the most critical components of a safe and healthy ferment. In most vegetable/fruit ferments, no starter culture is added. Rather, salt is added, which selects for the salt-tolerant lactic acid bacteria that are already present on the fruits and vegetables. These bacteria can tolerate very high concentrations of salt, whereas pathogenic bacteria cannot. The salt also draws moisture out of the vegetative matter, creating a natural brine in which the fruits and veggies ferment. The concentration range of salt to shoot for is generally between 2% and 4% salt by weight. You should not go below the lower end of the range, as this can open the door to pathogens. The higher end of the range, however, can be quite salty at the end of the day. Different concentrations can also select for certain lactic acid bacteria over others, producing different flavors in the final product. Earlier in my career, I tended toward the higher end of the range to ensure safe fermentations. Now that I'm a grizzled veteran, I tend toward a 2% salt concentration for most of my ferments. It depends on the ferment, though, and I will provide guidance in the recipes included in the book.

Since I primarily use vacuum-sealed bags for doing my lactic acid ferments (and I advise you to do so as well—see below), this makes determining the salt concentration about as straightforward as it gets. Just weigh the vegetable matter that is going to be fermented (including fermentable seasonings) and calculate 2% of that weight (i.e., multiply the mass of the materials to be fermented by 0.02). That would be the amount of salt to add for using a 2% salt concentration by weight. Obviously, if you are doing a 4% concentration by weight, you would take 4% of the measured mass of fermentables. If you are using a jar or crock for fermenting, which in some cases is necessary, such as when you're fermenting whole veggies like cucumbers, you will need to make your own brine. In these cases, tare your jar (or crock or whatever vessel you're using), which means put it on the scale and zero it with the container. If you do not have this functionality, just record the weight of the jar. Now, fill the jar with the materials to be fermented and then add your filtered water to the top of the jar. Record the mass of the materials to be fermented plus the water. Multiply that mass by 2%. This is the amount of salt you will add. I would not recommend adding any more than this, as it's actually quite a bit more than 2% of the weight of the vegetable matter. Pour the water out of the jar into a bowl and add the salt. Mix to dissolve the salt and then pour the salt brine back into the jar.

Now you're ready to ferment! For a more detailed discussion of salt use in fermentations, see the next chapter.

Now that you've sanitized your equipment and added the correct amount of salt, it's time to ferment. After the ferment kicks off, it's time to start checking for the tell-tale signs of a healthy, vigorous ferment (or one that has gone off the rails for one reason or another). Whether you're fermenting in a vacuum-sealed bag (I sure hope so—see below) or in another vessel, you should notice some bubbling within 24 hours. The vacuum-sealed bag will start to expand around this time, although it usually takes several days to fully expand to the point of bursting. I strongly advise checking the pH within 48 hours of the beginning of the ferment. The vast majority of ferments are acid based. The organic acids that are produced by the bacteria (primarily lactic acid if you're doing a lactic acid ferment) lower the pH to safe levels that protect the ferment from pathogenic bacteria. So, what is the pH threshold you should be looking for? A pH below 4.6 will protect your ferment from most pathogenic bacteria. This is another reason to ferment in a vacuum-sealed bag. Most pathogenic bacteria (or other microbes that will ruin your ferment, even if they're not going to kill you) are aerobic, meaning they need oxygen to metabolize and proliferate. By removing the oxygen in a vacuum-sealed bag, you remove this threat from your ferment. There is one particularly nasty microbe, however, that is anaerobic and the main reason you want to ensure that your pH is low enough. The bacteria in question is *Clostridium botulinum*, which may sound familiar, as it produces botulinum toxin, the most lethal substance known. Botulinum toxin has a human LD_{50} (the dose it takes to kill half the population tested in a prescribed time) of 1–3 nanograms (that's 1–3 billionths of a gram) of toxin per kilogram of body mass (considering the average male weight is about 70 kilograms, it would only take about 70 billionths of a gram to kill half the adult males tested) (Horowitz 2005). This is definitely not something you want to mess with. Fortunately, *Clostridium botulinum* doesn't like it too acidic and can't survive at a pH below 4.6. It shouldn't take very long to reach this pH, and most vegetable/fruit ferments will drop quite a bit lower than this value (into a pH range of 3.0–4.0). If your ferment is not there within 48 hours, you likely have a problem. You could let it go another day or two, but if it hasn't reached a pH of 4.5 by that point, it's not going to. Be safe and toss it!

Besides checking the pH, use your sensory organs to ensure that your ferment is safe. If it smells rank or rotten (provided you're not fermenting something super funky like natto), it probably is. Make sure to give your ferment a good whiff when you check the pH. Noses are incredibly sensitive organs that evolved to alert us to possible dangers, like protecting ourselves from consuming rotten food that will make us sick or kill us. As they say, the nose knows. So, use it! Our eyes can be pretty useful as well. The main things you're looking for are discolorations and yeast and mold growth. If you ferment in vacuum-sealed bags, you probably won't have these issues. When you cannot ferment in an oxygen free environment (or choose not to because you like to live life on the edge), you very well may experience these issues. At a minimum, you will probably grow some kahm yeast. This is a white, filmy substance that grows on top of the liquid. It's not harmful, but it doesn't taste good. This is the one that it's OK to scrape off and keep fermenting. If you notice any mold spots, I advise tossing it. You will inevitably see a ton of advice stating that it's OK to scrape it off and keep going. Or if it's white, then it's OK. But you cannot identify a mold species simply by the color of the mycelium (the fuzzy bits that we associate with mold). And many molds can express different colors depending on environmental conditions. Pathogenic molds produce mycotoxins (which is just a fancy word for fungal toxins) that are released into the solution on which the mold is growing. Since you don't know what kind of mold is growing on your ferment, you have no idea whether it's pathogenic and whether it has produced mycotoxins that are now in your ferment. These mycotoxins probably won't kill you (right away), but they may make you sick, and they can certainly build up in your system if you consume them regularly. I know it sucks to say goodbye to a ferment that you've been looking forward to, but your health is far more important. My advice, as a trained chemist and experienced fermentation scientist, is to throw it out if there is mold on it that you are not expecting. Now, there are certain exceptions. For example, sometimes when we're doing meat ferments and we're curing smoked sausages in the same cabinet as sausages intentionally inoculated with mold on the casings, mold will grow on the smoked sausage casings. If it looks like the *Penicillium* or *Aspergillus* mold that is intentionally growing on the other sausages, we'll wash it off with a little vinegar solution and keep on going. In general, if the mold is growing on a liquid solution (like the brine for your veggie ferment), throw it out. If it's growing on a solid substrate like a sausage casing or cheese, you may have a little more leeway. The mycotoxins cannot dissolve very readily into a solid substrate, so most will be concentrated in the layer with the mold. That said, if there is mold growing on sausage casings or cheese without a ready explanation for what it is and how it got there, my advice is to stay safe and dump it.

TOOLS OF THE TRADE

Since this book is about as nonspecific as it gets and covers virtually every imaginable type of fermentation, it would be impossible to cover all the tools necessary for each type. Plus, there are plenty of great books out there that are more specific regarding particular types of ferments that already cover the necessary tools. As such, I'll touch on some of the more critical tools to each type of ferment or tools that are used regardless of the type of ferment you are performing.

The use of the jar method versus the vacuum-sealed bag for lactic acid ferments is a hotly contested issue and one that I intend to settle once and for all (or at least provide my educated opinion, for what it's worth—the price of this book, I guess). So, let's make this easy. Stop fermenting in jars unless there's a good reason to do so (there are some, so keep reading). Get yourself a cheap vacuum sealer and some vacuum-sealing bags and use them for most of your vegetable ferments. I may sound a bit absolutist about this, but the vacuum-sealing method really works well. Sure, there are some ferments, like with whole vegetables (e.g., cucumbers) where it would be hard to ferment in a bag; but otherwise, leave the mason jars for storing your moonshine like a respectable southerner. Why am I so adamant about this? Because virtually every ferment that I've undertaken in a jar has grown mold or, at a minimum, kahm yeast, that white film that develops on the top, which isn't harmful, but which also doesn't taste particularly good. And every ferment I have ever done in a vacuum-sealed bag has fermented great with no mold or kahm yeast growth whatsoever. Why the stark difference? One word—oxygen! The lactic acid bacteria that are responsible for vegetable ferments are facultative anaerobes, which means they can tolerate a little oxygen, but they certainly don't need it to do their thing. There's no way to keep your mason jar oxygen free, regardless of how many rocks, weights, water-filled bags, or salt layers you may apply (although, in my experience, the water-filled bag is the best way to fill the headspace of a jar and prevent mold growth—but weights suck, so don't waste your money on them). A vacuum-sealed bag certainly isn't a perfect vacuum, but enough oxygen is removed to prevent any mold or yeast growth. But, you might protest, it's so hard to check the progress of the ferment if it's in a sealed bag. Well, that's kind of the point. Leave your ferment alone and let the microbes do their thing. You'll know that it's fermenting when the bag begins to expand from the gases produced during fermentation. And you'll have to snip a corner off the bag to check the pH and release the carbon dioxide gas occasionally (too much carbon dioxide will slow the bacterial metabolism and the ferment), and then reseal it again. One downside of the vacuum sealing technique, in my opinion, is the use of plastic over glass or ceramic. This is obviously a huge environmental and health issue, and if there were viable options other than plastic for vacuum sealing, I would use them. In the meantime, though, I would rather use plastic in these limited instances than run the risk of ruining a ferment and wasting food or potentially making myself or someone else sick. Another downside is the fact that fermentations may really do better in nonglazed ceramic containers than in other fermentation vessels. A study investigated the efficacy of glazed and nonglazed traditional kimchi ceramic pots, call onggi, relative to polyethylene, polypropylene, stainless steel, and glass, for kimchi fermentability, texture, taste, and health properties. The onggi in general outperformed the other containers, and the nonglazed onggi kimchi had the best overall lactic acid bacterial growth (and slowest aerobic bacterial growth, which is a good thing), texture, and taste as well as the greatest antioxidative and anti–cancer cell proliferation activity relative to the kimchi produced in the other containers (Jeong et al. 2011). A more recent study found that the reason for the onggi outperforming the other vessels was due to the permeability of the ceramic vessels allowing the carbon dioxide gas to escape, thereby lowering the gas levels to those more favored by the lactic acid bacteria, which explains why the nonglazed onggi outperformed the glazed onggi (Kim and Hu 2023). So, now that I've settled nothing, if you're doing a lactic veggie ferment, do it in unglazed ceramic containers (but, make sure to use your water-filled bag to prevent mold growth) or in vacuum sealed bags (and make sure to snip the corners and release that carbon dioxide regularly).

Now that we've covered what pH is and why it's important, let's measure it. The less expensive option is to use pH strips. They're sufficient for most applications, especially if you're mostly concerned with ensuring that your ferment is below the magic pH of 4.6; however, their accuracy is certainly limited. If you need a more accurate measurement, you could invest a bit of money to get a pH meter with a built-in probe. These will run you anywhere from about $20 (probably for a piece of junk) to around $100 for something a little more trustworthy. If you want to step it up even more, you could invest in a laboratory-grade pH meter with separate probes. These can run you anywhere from a couple hundred bucks to a few thousand. Using a pH meter, however, does not guarantee accuracy. They are only as accurate as the user. Unless you do, at minimum, a two-point calibration of your meter at pH 4.0 and 10.0 (even better is a three-point calibration, adding in pH 7.0 to the mix) before every nonconsecutive measurement, then your measurements are garbage. Additionally, the probes are very delicate and must be handled with care and stored in the buffer solution (or whatever solution is recommended for your probe) when not in use. All that said, even for the moderately priced meters with built-in probes, be prepared to buy new ones on a fairly regular basis. I've been using expensive, laboratory-grade meters as well as less expensive versions for decades, and I've been thoroughly disappointed with all of them. They inevitably malfunction on a regular basis and need to be replaced (for the separate meters and probes, it's usually the probe that goes bad, which is far less expensive than the meter, but still). So, you're sacrificing a little accuracy for a huge savings when using pH strips, and unless you're avidly monitoring pH and need the extra level of accuracy of a meter, I would stick with the strips. When measuring the pH, regardless of whether you're using strips or a meter, make sure never to stick the probe or strips directly into your ferment. Always take a little of the liquid from your ferment, put it in a secondary container, and measure the pH from there. Don't contaminate your ferment while trying to ensure that it's safe! Speaking of transferring some of the liquid from your ferment to measure the pH: disposable, plastic Pasteur pipettes are cheap and perfect for just such a job. If you don't want to use plastic, you can buy the glass ones for a little more money. You just need a separate bulb to fit on the end of these, as glass is a little less squeezable than plastic. The glass pipettes are disposable as well, but they can be cleaned and used multiple times if you're careful; and you can still recycle them when you don't feel like cleaning them anymore.

One of the most important tools for any kitchen, in my opinion, regardless of whether you're fermenting or not, is a nice scale. It can be analog or digital, although the digital ones are pretty reasonable these days, so I would recommend digital if you can do it. Make sure that it has a decent weight capacity (at least 2–3 kg, if not more like 5 kg) and that you have some different unit options on it, the most important being grams. Speaking of which, the fact that we don't use metric in this country still just absolutely baffles me. Not only is the metric system easier (everything's a factor of 10!), but you can use units that are appropriately small or large enough for any measurement. If kilograms are too large, you could switch to grams. If grams are too large, you could use milligrams. All we have are pounds and ounces. Pounds are often too large of a unit and ounces are not small enough. And it's perfectly sensible that there are 16 ounces in 1 pound. *That* makes the math super easy. Deep breaths, Brett. So, back to scales. Make sure that your scale, at a minimum, has grams. Now, why is a scale so important? Because gravimetric (by weight) measurements are far more accurate than volumetric measurements in general, but especially so when you are measuring dry ingredients. Why in the world would you measure flour or sugar with cups or teaspoons instead of by mass? I'll give you an example of how much of a difference it makes. I'm the cook in the family, but my wife generally does most of the baking. So, when pizza day rolls around (i.e., when I can convince my wife that it's time to make pizza), my wife is typically the one who makes the pizza dough. After experimenting with about a million different dough recipes, we settled on Roberta's Pizza Dough recipe from the *New York Times*. The recipe lists both volumetric and gravimetric measurements. When my wife followed the recipe, she would use the volumetric measurements, and I would inevitably complain that the dough wasn't flexible enough, making it difficult to shape into pizza rounds and, ultimately, making a dough that was not quite thin enough and a little too chewy. It was still quite good (I love you, honey!), but one time recently when my wife didn't have time to make the dough that day, I stepped in. I naturally used the gravimetric measurements in the recipe and, sure enough, the dough was pliable, easy to shape into rounds, and came out perfectly thin and crispy, with just the right amount of chewiness. My wife immediately wanted to know what I did differently. I told her I just followed the recipe but used the listed weights instead of the volumes. She has never used the volumes since. Now, of course that's just an anecdote, but think about it. Different flours will have different grinds, sugars and spices do, too. The same volume of a different grind would give you a completely different mass. That's why it doesn't make any sense to measure dry goods volumetrically, because you have no idea what grind was used on the ingredient listed. For liquids, that's not an issue, but a gravimetric measurement is still more accurate than a volumetric one, especially when it comes to the typical kitchen utensils that are used to make the volumetric measurements. The moral of the story is to buy a decent kitchen scale with a good weight capacity and different unit options that at least include grams. And use it!

A couple other items that are useful for a number of different ferments are hydrometers and refractometers. Hydrometers are thin glass tubes that are completely sealed, with a bulb at one end that is filled with lead or steel beads. They come with a cylinder that doubles as a case and a vessel for filling with the solution being measured. Once filled with the solution, the hydrometer is placed in the cylinder, bulb side down, and floats in the solution. They are labeled with graduations, the scale of which depends on the measurement they are designed for. The most common scale for hydrometers is specific gravity (also called relative density, which I will clarify momentarily), which is the ratio of the density of the solution being measured to the density of water at a specified temperature (which is 1.0 gram per milliliter at 4°C). Since it's a ratio of two density values, specific gravity is unitless. The hydrometer scale is set such that when it sits in pure water, the scale reads zero. The denser the solution becomes, the more the hydrometer floats, rising higher in the cylinder, with the scale set such that the specific gravity can be read at the line that is at the surface of the liquid. The specific gravity scale is great for alcoholic fermentations, as it can be used to measure the amount of sugar in the solution before fermentation. Instead of specific gravity, other scales that are common and useful for alcoholic fermentations are the Brix scale, the weight percent of sucrose in solution (useful for wine fermentations); and degrees Plato (°P), which is the same scale as Brix, but the weight percent of extract in solution (often used for beer ferments, as the pre-fermented wort is sugar plus other dissolved solids, such as proteins and starches). Another common hydrometer scale that is useful for our needs is a weight percent of salt, which is useful when making brines for lactic acid ferments.

Refractometers operate on a similar principle to hydrometers. These look similar to those toy periscopes you may have played with as a child. They are basically cylinders about 6 inches long and a couple inches in diameter with a viewing window in one end and an angled sample surface with shield on the other. A few drops of the sample to be measured are smeared on the sample surface and covered with the plastic shield. When you hold the cylinder up toward a light source and look through the window at the other end, you see the refractometer scale and a line created by your sample. Assuming the refractometer is calibrated, the position where the line falls tells you the value of your sample. Again, the scale could be specific gravity, Brix, °P, or percent salt, depending on what you are measuring. Refractometers operate based on the refractive index of the solution, or how much the solution will bend the light traveling through the solution. The higher the density of the solution, the greater its refractive index, and the more it will bend light traveling through it. So, when light shines through the solution on the sample surface, it bends a certain amount depending on the density of the solution. This shows up as the line of light viewed through the window that, on a calibrated scale, gives you the measured value.

Since hydrometers and refractometers are used for essentially the same measurement, you will probably use one or the other. So, which one to choose? Well, they both have their advantages and disadvantages. Hydrometers are generally cheaper, but nicer ones can certainly run as much as refractometers, which aren't that expensive either. Refractometers are easier to use and faster when making a measurement. You need only a few drops of sample to make a measurement with a refractometer, whereas you need at least 100 milliliters to make one with a hydrometer. Density decreases with increasing temperature, so you have to cool your sample to room temperature with either measurement device; however, a few drops in a thin film cools almost instantaneously, while a couple hundred milliliters in a cylinder takes a good bit. That's why it's faster and easier to make a measurement with a refractometer. Hydrometers should maintain their factory calibrations, as there are no moving parts on them. As long as the bulb doesn't break and spill some of the beads, then it should remain calibrated. Refractometers can lose their calibrations, but there is typically a calibration screw that can be turned to adjust the light line to the zero mark when measuring pure water. I generally prefer refractometers for the ease and speed of the measurement, but either one is fine. It just depends on your preference. When it comes to alcoholic fermentations, however, hydrometers and refractometers are both really limited to pre-fermentation measurements. Since they are both based on densities (or refractive indices, more specifically, for refractometers), this measurement becomes a lot more complicated after alcohol is produced. The density of ethanol is about 79% that of water, so once you have a solution with alcohol in it, the total density of the solution is dependent on both the alcohol and the water. The problem is, you don't know how much alcohol is in there, and that's what you're trying to measure. In other words, how much of the change in density is due to the sugars being metabolized during fermentation and how much is due to the alcohol that is produced from the metabolized sugars? Whether you're using a hydrometer or a refractometer, you really have no way to tell with these instruments. There are hydrometers that measure alcohol content specifically, but these are meant for finished products containing primarily alcohol and water, not the unfermented sugar that further muddles the measurement. To measure alcohol content, you really need more expensive instrumentation that isn't going to be accessible to the home fermenter. Otherwise, there are look-up tables that can be used to estimate your alcohol content based on your starting specific gravity and the measurement made post fermentation. These won't give you extremely accurate values, but they'll get you pretty close.

Finally, many (but not all) of the ferments covered in this book can be done at room temperature and require no additional equipment for controlling temperature and humidity. Some do require an environmentally controlled chamber for elevated temperatures and humidity. Fortunately, controlling temperature and humidity is not overly complicated or expensive and requires only a few additional pieces of equipment. The first is an old refrigerator or chest freezer. These work well because they're essentially insulated boxes. However, you won't be plugging in the appliance for many of the applications, and you could really get away with using an insulated box. For some applications, though (cheese and sausage, for example), you'll need the refrigerator or freezer for cooling rather than for heating. You'll also need a temperature and humidity controller. Inkbird has several versions that are available for less than $100. Then, a small space heater and humidifier can be used to heat and humidify the chamber, respectively. These are plugged directly into the controller, which is set to the desired temperature and relative humidity. The controller cycles them on and off to maintain the set values. If you were cooling the chamber, you would forego the heater and instead plug the refrigerator or freezer into the controller, which would cycle the appliance on and off to maintain the desired temperature. One issue with using these appliances as fermentation chambers is the drainage. When humidifying the chamber, moisture can build up and begin to condense on the inside. A small fan can be run inside the chamber to help keep the air moving, but the moisture has to go somewhere. If the appliance has a drain, keep it open with a small catch dish beneath the drain to capture the liquid that drains out. You will probably also need to wipe out the condensate that collects inside the chamber with a towel on a daily basis. Now you're all ready to begin on your fermentation journey!

Part I

FRUITS AND VEGETABLES

Chapter 1

PAST

——

Fermenting vegetables has been used as a means of preserving these perishable food items for thousands of years. Just add a little salt and let the lactic acid bacteria do their thing. Not only do the vegetables remain edible far longer after fermentation, but their nutritional content is also more accessible as a result of fermentation. Humans were able to grow such big, effective brains because we learned to cook our food, which is also the way we are able to sustain such nutritionally demanding brains. Cooking breaks down the macromolecules in foods, making the food matter more digestible in the limited human gastrointestinal tract. This allows us to consume far more nutritionally dense caloric content than we would otherwise be able to. Think of our close cousins, the gorillas. They spend about half their days eating. And by eating, I really mean chewing. Since they can't cook their food, they have to chew the plant matter they consume incessantly to break the recalcitrant vegetative matter down enough to be able to digest it and retrieve the nutritional content. That leaves them with very little time for innovation, artistic endeavors, and most importantly, social media. So, instead, they spend their time tromping around the jungle seeking more vegetative matter to chew on.

Fermentation achieves the same results as cooking. Instead of using heat to break down the food material, though, fermentation lets microbes do the heavy lifting. They effectively begin the digestion process outside of our guts so that we can finish the process inside our guts and extract maximum nutritional value from the vegetable matter we ingest. Fermentation takes time, though, ranging from a few days to weeks or more; that is way more time than we would ever want food hanging out in our guts. So, using fermentation we avoid any gastrointestinal distress while getting more nutritional reward than we would otherwise be able to, with the bonus of adding more good bacteria to our guts. On top of all that, we also get the added benefit of the healthful compounds produced during fermentation that we wouldn't normally get, such as organic acids like lactic and acetic acid. In fact, the authors of the External Fermentation Hypothesis posit that it was fermentation, rather than cooking, that led to the expansion of modern human brain sizes (by a factor of three over our early hominid ancestors) at the expense of our guts (which are fully a third smaller than other primates of a similar size) (Bryant, Hansen, and Hecht 2023).

CHOW-CHOW

Chow-chow, chow chow, or even chowchow, as you'll see it referred to, depending on where you are and who's spelling it, has the unique characteristic of being ubiquitous in households throughout the South (especially in grandma's house) and shrouded in mystery concerning its origins. I'll stick with chow-chow, because who doesn't love a little hyphenated redundancy.

Chow-chow also defies a simple description of what it comprises. It is kind of the southern US version of kimchi, a spicy, sour condiment that is a staple in many households, although versions of it are also found in Pennsylvania and elsewhere around the country as well as in the Maritimes of Canada. It usually has cabbage in it as well as any other vegetables one has lying around the kitchen or remaining in the garden before that first hard frost of the fall. Chow-chow is sometimes referred to as a southern relish, and it often draws comparisons to English piccalilli, which of course is derived from South Asian chutneys. Call it what you will, it falls into the general category of fermented vegetable (or at least sour, spicy, and maybe a little sweet) condiments that are found in homes throughout the world.

Versions of it stretching back a century or more often involved the addition of vinegar and sugar, so were not fermented, but quick pickled. This is how it is typically prepared nowadays as well. However, I'm going to go out on a limb and assume that the original versions were naturally fermented, thereby producing the organic acids that would preserve it for extended periods of time, much like the quick pickled version of today. There is scant historical evidence of fermented versions, mainly in the form of family recipes that have supposedly been passed down through generations. Regardless, we're going to roll with it. Even if original versions in this country weren't fermented, they should have been, and their international cousins certainly were, so there.

The name possibly derives from the French word for cabbage, chou. "Chouchou" is a French term of endearment roughly translated as darling or loved one. This etymological derivation of the term would seem to support the version of the origin story that chow-chow derives from the pesky Acadians of Novia Scotia in the eighteenth century who were expelled by the British during the French and Indian War for being, well, too French. Many of those who survived the expulsion ended up in Louisiana after the Spanish recruited them to the region to help stem the British encroach-

ment from the east. To the British, I say thank you for the roux, as this is what formed the foundation of creole cooking. Further support for this narrative is the fact that a version of chow-chow is found in Nova Scotia to this day. However, southern food historian John Egerton was fonder of the origin story that claims chow-chow derives from immigrant Chinese rail workers who brought with them their delicious pickled-vegetable traditions (Egerton 1993).

Regardless of its origins or your preferred use of hyphenation or word spacing, if you're from the South, you've probably had chow-chow. I've included a recipe for a fermented version of this delicious, spicy, tangy, sweet, pickled vegetable condiment. Serve it with hamburgers, hot dogs, pork, beans, corn bread, or pretty much anything you like.

WATAUGA KRAUT FACTORY

When you think of Western North Carolina, you probably think of the verdant, green Appalachian (correctly pronounced a-puh-LATCH-uhn, by the way) Mountains. Maybe Bluegrass music. But sauerkraut? Probably not.

But you would be wrong. My hometown of Boone, North Carolina, was for several decades the sauerkraut capital of the South. It all started in the early 1920s when a cooperative of Watauga County farmers and businesspeople started the Watauga Kraut Factory in order to do something with all the cabbage grown in the area. Even though the majority of people living in this region were of Scotch Irish descent, the eminently practical German tradition of preserving cabbage through fermentation apparently transcended any cultural barriers.

The cabbage itself came from many different sources but was probably largely the same variety, what is now referred to as Big Flat Head (I'll give you three guesses why this is the name). It's a large, specialty cabbage that produces many sweet, tender leaves per head, perfect for sauerkraut production, and that is still popular in this area to this day. It's difficult to establish with certainty whether this was the preferred cabbage, but based on photographs from the time—which show large, flat headed cabbages—and on accounts from old timers in the area, this seems to have been the case.

Not many crops grow as well as cabbage in the High Country of North Carolina, which has a shorter growing season than the rest of the region due to the elevation and which lacks much flat, arable land in general. Plus, it was and still is a staple crop that is nutritious and can be used in many different culinary applications. And the fermentation process unlocks even more of the nutrition. Cabbage is high in vitamin C and B vitamins and, as a cruciferous vegetable (cabbage, broccoli, brussels sprouts, etc.), it contains glucosinolates, compounds derived from glucose and amino acids. When glucosinolates break down, they release isothiocyanates, "mustard oils," which provide the bitter, pungent flavors associated with cruciferous vegetables. The release of isothiocyanates is a defense mechanism developed by these plants to ward off herbivorous insects (and young children, apparently). However, they are also potent bioactive molecules that exhibit antioxidant, anti-inflammatory, and antitumor activity. Fermentation of cabbage into sauerkraut is a very effective way of breaking down glucosinolates to isothiocyanates and releasing all that protective goodness. Now back to the Kraut Factory before we return to the science.

The Watauga Kraut Factory was opened in 1923 by H. Neal Blair, and in that same year the *Cleveland* (the county in North Carolina, not the city in Ohio) *Star*, a local periodical, reported, "Even though the workmen are new at this job, even now they are making about 15 tons of kraut per day," which translates to about 15,000 cans of kraut per day. From 1926 on, the factory was operated by the North State Canning Company under the direction of Dr. H. B. Perry and W. F. Miller (Jones House, n.d.).

The Kraut Factory was cranking out kraut until the early 1980s. By the 1970s, they were producing roughly 3,000 tons of kraut per year, or about 24,000 cases per day of that fine Watauga Chopped Kraut. The factory also canned sauerkraut juice and sold it as a health tonic to ward off colds, flu, and even hangovers. Apparently, there was even a drink referred to as the Watauga Cocktail, which included the kraut juice and tomato juice, a recipe that was developed with help from NC State. Just add some vodka and it sounds like a High Country Bloody Mary to me. I'll have to give that one a try.

What goes up must come down, as they say. Watauga County was becoming a more popular destination by the 1970s, with the growth of Appalachian State University and, well, paved roads up to the High Country. Apparently, the entire state of Florida realized around this time that they could escape the swampy heat of their state for the cool mountains of North Carolina. This resulted in much of the farmland being sold off for second-home development. That, coupled with a devastating fungal disease that impacted the cabbage crops at the time, resulted in a steep decline in sauerkraut production by the late 1970s. Ultimately, the factory shuttered its doors by the early 1980s (McDaniel 2015).

Even to this day, decades after the factory closed, the memory lingers on. Many locals in Boone refer to the creek that runs through town and the campus of Appalachian State University, which is officially and uncreatively named Boone Creek, as Kraut Creek. The name derives from the fact that when the factory was in full swing, waste sauerkraut effluent would be dumped in the creek, giving its waters a somewhat tangy aroma reminiscent of, well, kraut. This has had a serious impact on the native trout population, as apparently, they have been spotted wearing lederhosen, which is particularly difficult considering they don't have legs.

Now, back to some science. Sauerkraut is one of the easiest vegetable ferments to prepare. It's just cabbage, salt, and time. That said, the devil is in the details. Typically, the cabbage is soaked in brine or dry salted to 2% by weight and fermented for about 3 weeks at room temperature. The lower salt concentration and lower temperature favor the heterofermentative bacteria, *Leuconostoc mesenteroides* and *Lactobacillus brevis*. These are lactic acid bacteria, the typical food fermenters that produce, shocker, lactic acid, which gives that tangy sourness to yogurt, pickles, and sauerkraut. However, since they are heterofermentative, they produce acetic acid, ethanol, and CO_2 as well as lactic acid, the combination providing a little depth of flavor. After a week or so, the brine becomes too acidic for the heterofermentative organisms, so the homofermentative bacteria, *Pediococcus cerevisiae* and *Lactobacillus plantarum*, which produce mostly lactic acid, take over and finish off the fermentation. The homofermentative bacteria can also be selected for by using a higher salt concentration (about 3.5%) and higher temperatures (greater than 30°C) if purely lactic acid is what you're shooting for.

Chapter 2

BACKGROUND

―

Not only has fermenting vegetables and fruits been the primary means of preserving perishable food items for millennia, it has also expanded food options that otherwise wouldn't be accessible to humans. The microbes, primarily lactic acid bacteria in the case of vegetable fermentations, produce organic acids (mostly lactic acid in this case), which provide some antimicrobial protection themselves, and lower the pH to a level that protects the food items from harmful bacteria. As the beneficial microbes consume nutrients during fermentation, they are further protecting the food against pathogenic bacteria that no longer have those necessary nutrients available to them.

Foods that are fermented are generally easier for humans to digest. As omnivores, we don't have overly complex gastrointestinal systems capable of easily digesting raw foods. As such, we must either cook our food to partially degrade it and make it more nutritionally accessible to us or we can ferment it, which serves the same purpose. It's either that, or spend the bulk of our days chewing, as our cousins the gorillas do.

Not only are fermented foods easier to digest, which inherently makes them healthier, the fermentation process also produces compounds—such as organic acids like acetic, gluconic, and glucuronic acids—that have proven health benefits. Not to mention the probiotics you get from live, fermented foods. There is an argument for doing your own ferments too, as you are encouraging the beneficial bacteria found in your immediate environment. It is likely more advantageous to encourage the bacteria that are already part of your ecosystem than to take probiotic supplements with a random assortment of bacteria that may not otherwise be present in your environment or even to consume live cultures in fermented foods that were prepared thousands of miles away. That's not to say that probiotics are bad, whether in supplement form or in commercially purchased food items. It's just that the benefits may be greater if you're working with bacteria that already call your gut home.

Then there are the foods that we wouldn't be able to eat at all unless they are processed in some manner to degrade the naturally occurring toxic compounds found in them. For such foods, like cassava, a staple in many parts of Latin America, Africa, and Asia, fermentation is necessary to degrade the toxic cyanide compounds, because boiling and drying alone may not do the trick (Halake and Chinthapalli 2020). Can you imagine how the first people to ferment cassava figured out the process? Was it accidental? That is likely, given that most of our first fermentation forays were probably lucky accidents. How many people fell ill or died before the process was discovered, though? And cassava is one thing. What about *hákarl*, fermented Greenland shark? Who figured that one out? Apparently, some very hungry Icelanders back in the day. Anyway, I think we've only scratched the surface on using fermentation to access food products that are currently unavailable due to their toxicity. See chapter 4 for more on that thought.

Fermentation is also a low-energy way to preserve foods. No electricity is required! As we like to say, fermentation is the original sustainable practice. And, oh yeah, fermented foods taste great! Fermentation generates tasty compounds that wouldn't otherwise be produced, like the aforementioned organic acids. But also amino acids, which lead to the umami taste we all know and love, described as savory and associated with foods like well-aged Parmesan, mushrooms, and beef broth. That's right, proteins are made of amino acids. When the proteins are degraded to their individual amino acids during fermentation, this can lead to the umami character. In particular, when the amino acid glutamate is combined with certain nucleotides and the minerals sodium and potassium, this leads to the umami taste.

You may be thinking to yourself, where have I heard that name glutamate before? Oh yes, from the nefarious monosodium glutamate, or MSG, that terribly toxic compound that gives me horrible migraines. Except that it doesn't. And there's absolutely nothing wrong with MSG, unless of course you don't like umami character. It was the subject of an unfortunate association of the compound with feeling unwell after eating large meals at Chinese restaurants, beginning in the 1960s, that came to be known as the Chinese Restaurant Syndrome. But the early scientific studies aimed at establishing any health impacts from consuming MSG were flawed to the point of making them meritless. More recent, less flawed studies have shown no association of the compound with negative health consequences (Levinovitz 2015).

Taste and Flavor

Taste and flavor, despite being used interchangeably in lay terminology, are not the same thing. First, the easy one. A human is capable of detecting five tastes: sweet, sour, salty, bitter, and umami. That's it, although there is an argument for a sixth taste: metallic. To date, though, metallic is left on the outside looking in. These tastes are detected by the taste buds on your tongue, the tiny chemical receptors that respond to compounds that produce these tastes. And, yes, there are different types of taste buds that detect different compounds, but all the taste buds are found on the same regions of the tongue. That is, different regions of the tongue do not respond to different tastes, as was the common thinking several decades ago based on a flawed scientific study. There are just certain regions of the tongue where the taste receptors are concentrated. The evolution of these taste receptors likely happened a long time ago, as they are a primary means of detecting foods that are (or are not) safe to eat. Sweet means sugar, or an easy source of energy. Sour indicates unripe fruit or fermenting (or rotting) food, so proceed with caution. Likewise, bitterness can indicate toxicity,

so a no-go. Salty directs us to much needed minerals, and umami means muscle-building proteins.

Flavor is the more complicated one. It's actually a combination of taste, aroma, and trigeminal sensations. We've covered taste. Trigeminal sensations are the result of the sensory feedback we receive from the trigeminal nerve, the largest of the cranial nerves that provides sensations in the face. In this case, the sensory feedback provided by the trigeminal nerve that contributes to the overall flavor sensation includes astringency, spiciness, and hot and cold. Aroma is the combination of orthonasal olfaction, which simply put is smelling through the nose, and retronasal olfaction, which is mouth smelling. This is a little misleading, however, as the olfactory bulb, a concentration of nerve receptor cells that detect aroma compounds, is located at the top of the nasal cavity. As you're eating things, which most people tend to do with their mouths, the volatile compounds in the food travel to the back of the throat and then up through the nasal cavity to the olfactory bulb. Whereas taste buds and the trigeminal nerve are pretty limited in what they can detect, the olfactory receptors can detect thousands of different aroma active compounds. And they're extremely sensitive. We have instruments in our lab that cost as much as my house (and I have a theory that the more expensive an instrument, the less likely it is to work properly) to detect these same compounds that are not nearly as sensitive as the human nose. Granted, different people have different sensitivities to different compounds, but in general we have yet to invent an instrument as sensitive as the nose. This is no evolutionary accident either. Just like our tastes allow us to identify good or harmful foods, smells alert us to the presence of tasty treats (think fruity smells or roasting meat) or danger (I smell a tiger or that meat smells a bit off) just like our taste buds. When somebody loses their sense of smell (which happens somewhat frequently as a result of COVID-19, for example), they can still detect the five tastes as well as the trigeminal sensations. So, if someone who lost

their sense of smell were eating their favorite lamb vindaloo, they would still get the spice and temperature of the dish as well as the saltiness, maybe a little sour from the tomatoes, and probably some umami. But that's it. You can probably see why smell is such an important contributor not only to flavor but also to our overall health and well-being. What if the vindaloo had been sitting out for several hours without your knowledge? Sure, maybe you'd get a little more sour taste as a result, but any indications of rot would be lost. Yikes! Also, it would just make eating a lot less interesting and enjoyable, which is why many people who lose their sense of smell end up losing weight as well, since they lose their desire for a lot of foods that they used to really enjoy. So, the next time you're eating your favorite dish, think about all of the factors that go into the flavor of the dish that make it so good. Or, better yet, just enjoy it! •

SALT

Arguably the most important ingredient for a good vegetable ferment, besides the veggies, of course, is salt. Most nonscientific literature will advise against the use of iodized salt for fermentations because of the widely held belief that the iodine in iodized salt will inhibit microbial activity. However, according to a recent study (Müller et al. 2018), the use of iodized salt in sauerkraut fermentations found that there was no significant impact on the fermentation performance of microbial populations. And just so you know, the fermentations in these studies were done both with starter cultures (i.e., specific lactic acid bacteria cultures were added to the cabbage to initiate the fermentation) and without starter cultures (i.e., no microbes were added to the cabbage, just the salt to select for the good lactic acid bacteria). Additionally, in the fermentations performed without a starter culture, which is the typical method used outside select industrial applications, the iodine did appear to inhibit yeast and mold populations, unwelcome and potentially harmful interlopers in this particular fermentation. That said, it should be noted that the study used a 1% salt concentration, which is on the very low end of salt concentrations for any vegetable ferment. Plus, the study investigated only sauerkraut production, so to extrapolate beyond that particular fermentation is not advisable.

Even though using iodized salt may be less of an issue than many believe, I prefer noniodized sea salt or kosher salt for my fermentations. They're no more expensive than iodized salts, plus the iodine can brown the vegetables in the ferment. You could use some ultra-fancy, expensive salts as well, such as Himalayan pink salt or *fleur de sel*, but remember these salts tend to have a high mineral content. The mineral content could add some flavor to the final product, but at high enough concentrations it could also inhibit microbial activity. Iron impurities, such as those found in some salts, can blacken the vegetables. Magnesium impurities can cause bitterness, while carbonates can result in a softening of the vegetables.

The salt itself inhibits microbial activity, at least that of the microbes you don't want in your ferment. Lactic acid bacteria, the "good bacteria" that you're selecting for with the high salt concentrations, are halophilic, or salt loving (more like salt tolerating, really, at the concentrations used for fermentations), whereas the "bad bacteria" that can make you very ill or at the very least ruin your ferment, are not. There will be hundreds of microbial species present on the vegetable matter, including bacteria, yeasts, and filamentous fungi (molds). The idea is to create an environment that is conducive to the ones that will successfully and efficiently ferment your vegetables, the lactic acid bacteria. The specific lactic acid bacteria that will likely be doing the heavy lifting of the fermentation are *Lactobacillus plantarum, Lactococcus lactis, Lactobacillus curvatus, Pediococcus acidilactici,* and *Lactobacillus brevis,* or some combination thereof, along with many other potential players (González-Quijano et al. 2014; Watts et al. 2018). Lactic acid bacteria, you may have already guessed, produce lactic acid during fermentation along with many other metabolic byproducts. The bacteria can be homofermentative, that is, produce primarily lactic acid; or they can be heterofermentative and produce acetic acid and carbon dioxide along with lactic acid as the primary metabolic end products of fermentation. Some lactic acid bacteria are primarily one or the other, while many others alternate depending on environmental conditions. The lactic acid (and possibly acetic acid) that the bacteria produce drops the pH of the solution, thereby further preventing harmful bacteria from taking hold. Finally, as the lactic acid bacteria use the available nutrients for their metabolic needs, they deplete the environment of nutrients that harmful bacteria could otherwise potentially use. So, once you add an adequate amount of salt to the vegetable matter and select for the lactic acid bacteria, they will take care of the rest.

In addition to inducing favorable microbial activity, salt allows water and sugars, which are used by the fermenting organisms, to be pulled from the vegetables. The salt also makes crisper vegetables by hardening the plant pectins (polysaccharides found in plant cell walls) and decreasing the activity of pectinase, an enzyme that degrades pectins and ultimately makes vegetables mushy. Last, but certainly not least, salt adds flavor to the final product, just like with other food products.

Eek, My Garlic Is Blue!

Don't be surprised if your garlic turns a blue-green color during fermentation. That's just a number of enzymatic and nonenzymatic chemical reactions (enzymes are proteins that act as catalysts to accelerate chemical reactions) involving sulfur-containing amino acids and resulting in colored compounds that appear to be unique to garlic but that are, in fact, structurally similar to chlorophyll compounds (Imai et al. 2006), hence the greenish color. These compounds don't impact the flavor of the garlic, just the color. The same reactions occur when the garlic is damaged. This is a clue as to the nature of the reactions that ensue. The color change is not due directly to the reduction in pH during fermentation. The pH does, however, dictate which reactions occur, with the nonenzymatic ones favored at lower pH ranges (below 6, which is definitely the range your ferment should be in!) and the enzymatic reactions favored at a pH above 6 (Bai, Chen, et al. 2005). All enzymes have a preferred pH range in which they operate. Outside of those ranges the enzymes start to lose their shape due to changes in the charges on the molecule. The reactions in this case, whether assisted by enzymes or non-enzymatically assisted, occur simply from a mixing together of the compounds that occur following damage to the garlic cells. The organic acids produced during fermentation, namely lactic and acetic acids, damage the garlic cells, thereby releasing the chemical compounds involved in the reactions into solution and allowing the colors to be produced (Bai, Li, et al. 2006). •

A QUICK WORD ABOUT VEGGIES

The vegetables that you use should be organic or at least as free of pesticides and fungicides as possible, as these chemicals can kill off the native bacterial populations on the vegetables. Before use, wash the vegetables well with water, but not with any bleach or other antimicrobial product, which could harm the good bacteria. If you're on a municipal water supply, you should use a filtered source that removes the chlorine, as that could also inhibit fermentation. Also, locally sourced vegetables will more likely have a similar microbial population to those found in your gut, which will help to promote those good bacteria and prevent gastrointestinal distress from disrupting those populations.

Chapter 3

PRESENT

——

CURRENT FERMENTED VEGGIESCAPE

Go to any grocery store these days, not even the fancy, hippy ones, and you see fermented vegetable products everywhere, from sauerkraut to kimchi to pickles to hot sauces, and lots in between. And some of them are from large, nationwide brands; but if they're live fermented products, they are generally not shelf stable without refrigeration, so many companies stick to more regional distribution. Which means you'll find more local fermented vegetable products in a store near you. One company that has actually been around for quite some time, making products that were certainly not popular in this country until much more recently, is the American Miso Company. Located in tiny Rutherfordton, North Carolina, American Miso was founded in 1979, when most people in this country probably had no idea what miso even was. It happens to be the largest producer of organic miso in the world, selling over a million pounds per year. John Belleme, the founder of the company was a medical researcher in Miami in the 1970s who became interested in the macrobiotic diet, a low-fat, high-fiber eating plan that focuses on whole grains and vegetables, adopted from Zen Buddhism and developed by Japanese philosopher George Ohsawa. The macrobiotics movement in this country was centered in Boston at the time, so John traveled to Boston where he met his future wife and cofounder of American Miso, Jan. There, they learned about miso and were so inspired by its healthfulness, they ultimately put together a group of investors to start the American Miso Company. First, though, they had to learn how to make it. To do so, they traveled to the Togishi prefecture in Japan where they lived with and learned from Takamichi Onozaki, a fourth-generation miso master. They spent 8 months living in his traditional home with no running water or central heating and learning the trade. Upon returning to the United States, they needed to figure out where to locate their production facility. Since their investors were from Boston and Miami, they chose a middle ground to meet and discuss options. That middle ground happened to be Rutherfordton, which they quickly realized had a similar climate to the prefecture in Japan where they learned their craft as well as an abundant, clean water supply. As luck would have it, there was a 120-acre tract of land that was available for sale in Rutherfordton, and the American Miso Company was officially born. Their first batch, traditional red miso, was fermented for a year and available for sale in 1981. Before selling it, however, they needed to make sure it was up to snuff. They got the master himself, Takamichi Onozaki, to come to Rutherfordton, where he sampled the batch and gave them his blessing. The Miso Master brand, with several different options nowadays and distributed by Great Eastern Sun Trading Company, has been selling products ever since.

Another product that hasn't been around for nearly as long but that is equally as unexpected is Bluegrass Soy Sauce made by Bourbon Barrel Foods in Louisville, Kentucky. Bourbon Barrel Foods is a company that specializes in food products that celebrate Kentucky's exceptional bourbon tradition. Their original product, Bluegrass Soy Sauce is the only micro-brewed soy sauce produced in the country and the only soy sauce in the world that is fermented in Kentucky bourbon barrels. Since the barrels can be used only once by the bourbon industry, founder Matt Jamie figured why not make soy sauce in the used barrels. After learning how to make it by researching the process online and through plenty of trial and error, he dialed in a recipe that has received plenty of praise from chefs and customers alike. Using soybeans, soft red winter wheat, sea salt, and limestone filtered water, the mixture ferments for a year in the barrels, after which it is pressed, and the resulting liquid soy sauce is bottled. With notes of bourbon and toasted oak flavors from the barrels, it's worth picking up a bottle and splashing some into your next recipe.

FERMENTED HOT SAUCES

As a professor at Appalachian State University who started not long after a certain infamous marketing video for the university was released, I would be remiss if I didn't include a fermented product that was HOT HOT HOT (www.youtube.com/watch?v=pVENWl8uBeg). OK, I'm done with the puns ... for now. So, I'll just cool it and focus on something that's been burning up internet forums and palates alike for several years, but less so supermarket shelves. There are products out there that are made using fermentation as part of the production process. Two of the more popular ones are, of course, Tabasco, produced for well over 100 years in the Southeast, on Avery Island, Louisiana; and Sriracha, which has recently taken the world by storm. In both Tabasco and Sriracha production, however, vinegar is added, presumably for flavor enhancement and to stabilize the products at a low pH, but which also dilutes the organic acids and other metabolites produced through fermentation. Don't get me wrong, I'm a fan of both Tabasco and Sriracha, but these hot sauces are not purely the end products of fermentation. In addition to the vinegar that is added, which overshadows the lactic acid from fermentation, they are both heat stabilized, thereby killing any microbes remaining from the fermentation process. They add a spicy bit of flavor to plenty of dishes, but you will not be getting your probiotics with either product. This section focuses on living hot sauces, the direct result of fermentation, with live microbes still present. This is probably why there aren't that many on store shelves, as they must be kept refrigerated and still have a limited shelf life.

So, what does a purely fermented hot sauce entail? Well, hot peppers of course. And there are plenty to choose from. Everything from mild and flavorful to hot and flavorful or just plain hot, with virtually every flavor and heat combination possible in between. The Southeast is also home to one of the hottest peppers on the planet, the unforgivingly, absurdly, over the top, tongue-shredding Carolina Reaper, a pepper that also happens to be quite flavorful. I've made a few fermented hot sauces using the Reaper that were both tasty and hot, really hot.

There's more to fermented hot sauce than just hot peppers, though. In addition to the peppers, add in any other vegetable or fruit matter, herbs, and spices you see fit. Onion and garlic can add a nice foundation of flavor, just like in other dishes. Carrot adds a nice flavor to the ferment as well, but it takes a bit longer to break down than the rest of the vegetable matter.

Thus, adding some roasted carrot after fermentation can add some pleasant earthy notes to the hot sauce (and it can reduce the overall spiciness if you unintentionally went a little overboard). Similarly, you may want to hold off on the herbs and spices until after the fermentation. Not only can some of the flavors become muted from the fermentation process (e.g., cilantro, the flavor of which is typically lost during fermentation), but you can also adjust the flavor of the final product more to your liking by adding them after the ferment rather than before and hoping for the best.

Chapter 4

FUTURE

——

Predicting the future is uncertain work. I think that's probably why you don't see many oracles around these days. It's just too stressful. I mean, they would get absolutely destroyed by internet trolls. However, because I've never been overly committed to doing the sensible thing, I'm going to dedicate these "Future" chapters to doing just that—trying to predict the futures of veggie, meat, dairy, and beverage fermentations. OK, well, maybe predicting is a strong word. I'm going to do some educated guessing about what the future may hold for these types of fermentations. And to be completely honest, most will be less prognostications than a brief coverage of work that is already starting to happen in certain fields and that may (or may not) become more important down the road. Now that I've thoroughly exhausted all excuses for when my predictions do not pan out, let's get started.

ALTERNATIVE PROTEINS

If there is one predicted future scenario for fermented vegetables that I feel most confident will indeed come to fruition, it's fermentation-derived alternative proteins. I certainly hope that it does become as important as I think it will, not only for the health of the planet and as a means for feeding an ever-growing population but also because we're beginning research in this field in my lab right now. Before I explain what fermentation-derived alternative proteins are, let's start with a little background into the issue.

In order to meet growing global demand from an increasing global population that is both growing more affluent overall (i.e., eating more animal proteins) and projected to reach 10 billion people by 2050, meat production would have to double from its current rate. However, even at the current rate of meat production, the climate impacts from methane emissions and the environmental impacts from land use and decreased biodiversity are immense. Doubling meat production would be devastating and prevent many countries from meeting their future emissions goals. At the same time, people have to eat. And proteins are a necessary part of any healthy diet. So, clearly things will have to change. Animal proteins are not going to be able to supply the global population's dietary needs, regardless of the climate and environmental impacts. Enter alternative proteins.

There are three main classifications of alternative proteins—cultivated meats, plant-based "meats" and other "animal" products, and fermentation-generated alternative proteins. Cultivated, or cultured, meats are the only one of the three that are actually identical to the same meats that are grown by, well, the animals themselves. Since they are actually still animal protein (although devoid of any animal), I'll cover this form of alternative protein in part II, "Meat," so hold tight for that.

Plant-based meats, despite the oxymoronic name, are the ones that you're probably already familiar with, whether you know it or not. Have you heard of or tried Impossible Burger or Beyond Burger? These products are a type of biomimicry if you will. They are completely plant-derived products (no animals harmed in their making, at least not directly) that are meant to look and taste like their animal-derived meaty counterparts. They do this by incorporating plant derived fats, proteins, and other minor components that are similar to the animal derived compounds found in the muscle tissue into a package that looks, feels, and tastes (somewhat) just like the animal derived product. As far as the alternative protein universe is concerned, plant-based meats are by far the major player currently. Many fast-food chains are carrying them, and even some big players like Tyson and Nestlé now have plant-based meat products as well. Although the majority of the plant-based meats are derived from plants, hence the name, many also incorporate some fermentation-derived compounds. For example, myoglobin, a protein found in animal muscle tissue, gives the meat derived from this tissue its signature flavor (and color when curing meats). Myoglobin can also be derived through animal-free fermentations and then incorporated into plant-based meats to give them a more similar flavor to the animal-derived meats.

Speaking of which, fermentation-derived alternative proteins are the third pillar of the alternative protein world, arguably the most nascent of the three and, in my opinion, the one with the greatest potential upside for two main reasons. First, there are so many options for fermentation-generated alternative proteins, which I'll touch on momentarily. And second, they're generally inexpensive to produce, or will be once there are enough production facilities online. The reason for the wealth of options is that there are multiple fermentation methods, microbes, and substrates that can be used to derive the alternative proteins. Traditional fermentation has been used for centuries to convert raw foods into products that are not only delicious but also more nutritionally accessible by the human digestive system. And fermentation is used to produce alternative proteins. Think tempeh, for example. The *Rhizopus* fungus converts soybeans into umami-rich tempeh. Similarly, lactic acid bacteria convert raw dairy into cheese and yogurt. These are technically alternative proteins that are being produced; but granted, that's a bit of a stretch, because what we're talking about here are more novel protein sources than the traditional options that have been available for ages. Still, traditional fermentation can be used as a source of novel alternative protein products by improving the flavor and nutritional qualities of raw materials through the fermentation process.

Biomass fermentation makes use of the microbial biomass itself as a source of alternative proteins. Microbes grow very rapidly and can produce multiple generations of new cells through asexual reproduction within a day. This method of growing microbial biomass as a source of alternative proteins presents the most scalable option because of the microbes' rapid growth cycle. A number of different types of microbes are being studied for their potential applications in biomass fermentation, including yeasts, molds, and microalgae. Currently, microalgae seem to be the preferred microbe, and they are being grown heterotrophically for this purpose, which is a fancy way of saying that they are not exposed to sunlight to produce their own carbon but are instead being fed sugar as their source of carbon. I imagine this is a more consistent method to grow and fatten them up than relying on sunlight to do the job. In general, the intact biomass thus derived can be used as an alternative protein source, or the biomass can be processed to some degree, which often just means that the cells are lysed so that the microbes spill their guts and the proteins can be concentrated further. Most of the microbial biomass now grown is currently being used as an ingredient in an alternative protein product rather than as the product itself, but that may change as new microbes are investigated and new products are developed.

The final example of alternative protein production using fermentation involves precision fermentation. This is when the microbes are used as teeny tiny factories to crank out specific ingredients that can be added to other products. Obviously, since we're covering alternative proteins, microbes are often used to produce proteins, and many microbes can be genetically modified to produce egg and dairy proteins as well as those proteins that are normally found in meat, like myoglobin and heme proteins (e.g., hemoglobin). However, they can also be used to produce other functional ingredients such as vitamins and fats. These ingredients can then be used in plant-based products to give them better flavor, texture, and consumer acceptability in general.

We are still at the beginning stages of the alternative protein landscape, but more companies are starting and more products are becoming available every year. You can now get mycelium (the "root-like" stringy or fuzzy structure of a fungus that sometimes produces fruiting bodies like mushrooms) steak and bacon or biomass protein sausages to throw on the grill. Heck, you can get a foie gras now that is goose free. It's made from mycelium-derived proteins. How about sliceable deli "meat"? Yup, got that covered, too, with animal-free meats made from koji mycelium. And that's just some of the animal-free meats. I have to admit that I haven't tried most of these products, as they are so new, and many are being produced on the West Coast with limited or no distribution to the South as yet. Or they are simply too expensive for my public servant salary. As the technology continues to improve and more producers get into the market, though, competition will undoubtedly drive the prices down. More competition means better overall quality as well, with flavors and textures that either rival animal products or are simply delicious, protein-rich products that don't even need to imitate animal products. And the potential climate impact of these products is enormous. A recent modeling study out of Germany published in the journal *Nature* projects that by replacing just 20% of ruminant meat consumption globally by 2050 with microbially derived proteins would reduce carbon dioxide emissions by half, while also reducing methane emissions and preventing further deforestation for pastureland (Humpenoder et al. 2022). Like I wrote above, fermentation-derived alternative proteins are one of the future scenarios that I feel most confident will, indeed, come to fruition. I'm certainly excited to see what comes from this developing area and, more importantly, to try a lot of these products!

DETOXIFYING RAW INGREDIENTS

What else am I seeing as I gaze into my crystal ball of the future of fermentation? Well, this is not necessarily something that I see as a big wave for the future of fermentation, but it is one that is interesting, nonetheless, and that I plan to focus on in my lab moving forward. We already use fermentation as a means of detoxifying raw ingredients and making them edible. Think cassava, the tuber that serves as a major food source for hundreds of millions of people worldwide but that also contains toxic cyanogenic glucosides (think cyanide) in the raw plant. Fermentation is a critical component of degrading its toxic compounds to make it suitable for human consumption. Well, what's a toxic plant in the South that everyone is familiar with? If you said, "Pokeweed," you win the prize. Every part of the pokeweed plant is toxic (potentially lethally so), and yet pokeweed sallet (the sallet—or salat or sallat—spelling apparently indicates that it's cooked) is commonly eaten throughout the South. OK, commonly eaten might be a bit of a stretch. But it is eaten throughout the South, primarily by older generations who may have grown up eating it. To prepare pokeweed sallet, the young greens that emerge in the spring (another reason why it was probably consumed back in the day, as a survival food when the stored reserves of food were almost depleted in the spring and the planted crops hadn't yet emerged) are harvested and then boiled three times, with the water discarded after every boil. The boiling degrades the heat labile toxins in the greens or at least leaches them into the water so that they can then be removed along with the discarded water. Then the greens are sautéed (traditionally in bacon fat) and served, possibly with a little vinegar-based sauce or dressing. Apparently, it's not too bad either, with a flavor reminiscent of cooked spinach or mustard greens with maybe a hint of asparagus. Admittedly, as you could probably tell, I have not tried pokeweed sallet, despite the fact that it grows everywhere in my yard. Yes, I will eat fermented Greenland shark, but pokeweed does scare me a bit. It is quite toxic, and though consuming a bit of it raw may not kill you, apparently, you will wish that you were dead from the vomiting and diarrhea that will ensue. Plus, the last thing I need is more greens in the spring, considering that our CSA is loaded with every leafy green imaginable until the weather warms up a bit and the less cold tolerant crops begin to arrive. So, survival greens are not generally top of my list. Especially ones that I have to boil multiple times before it is safe to consume them.

The question I want to answer is, Do you have to boil them? Or could fermentation possibly effect the same result, degradation of the toxins? It should, in theory, be possible. The primary toxin is *phytolaccotoxin* (which is not remotely helpful in the identification of the toxin, as *Phytolacca* is the genus of the plant), which is a saponin, a class of compounds sometimes referred to as triterpene glycosides (which is a bit more descriptive) that are large, typically bitter tasting, often toxic molecules produced in many plants. The key reason for my hypothesis that these are potentially fermentable is the presence of glycosides, sugar molecules that hang off the backbone of the molecules. There are bacteria and fungi that are known to ferment saponins. It's just a matter of whether they will ferment these particular saponins. The other catch is that there are many other toxic compounds found in pokeweed, some of which probably have not been identified. What on the surface may seem like a fairly straightforward question to answer will, no doubt, turn into a months-long endeavor in frustration. Ah, the joys of research.

The fun doesn't stop there, though. A favorite pastime of mine in the warm months is strolling through the vast forests of southern Appalachia foraging for mushrooms (or tending my ever-growing mushroom garden). Despite our tendency in this country to slice them raw on top of salads, most mushrooms should generally not be eaten raw. This is mainly because they are largely composed of chitin, the same polymer that makes up the

exoskeletons of insects and crustaceans. It is generally regarded as inadvisable to consume the lobster shell when you are treating yourself to a lobster dinner, as the shell is not the most digestible part of the animal. Along those same lines, raw mushrooms, due to the chitin content (although, admittedly, nowhere near as much as a lobster shell), are very difficult to digest and can cause gastric distress if consumed in quantity. That's why you should always cook your shrooms. The heat breaks down the chitin and makes them much more digestible. That said, there are some mushrooms that, beyond the chitin content, need to be cooked well in order to break down some heat labile toxins. I'm not talking about toxins that can kill you, like the ones found in some species of *Amanita* or *Gallerina*. For those, I advise very strongly to stay far away from them (although there are some *Amanita* species that are edible and quite delicious). Rather, these are toxins that will make your gut fairly upset for a while if the mushrooms are not properly cooked. Nothing that you want to go through, but nothing that will kill you either. The mushrooms I'm referring to specifically (although I'm sure there are many more) are chicken of the woods, *Laetiporus* sp., and honey mushrooms, *Armillaria* and *Desarmillaria* sp. (side note—*Armillaria ostoyae* is in contention for the largest organism on the planet, where an individual fungal organism, fittingly referred to as the humungous fungus, in the Malheur National Forest in Oregon, covers something like 9 square kilometers with its underground mycelium and is likely thousands of years old). Rather than cooking, however, could you ferment the mushrooms to degrade the heat labile toxins? Apparently, it's quite popular in Eastern Europe to pickle honey mushrooms, although this is typically done after the mushrooms have been boiled. The problem with researching the efficacy of fermentation in degrading the toxins is that the toxins in question are not identified. Oh, and then there's the fact that the same species of mushroom foraged in different locations could have very different levels of toxicity. Not to mention the fact that some people seem more susceptible to the undercooked mushrooms than others, so there might be some type of allergic reaction causing the response. There are a number of problems related to this possible area of research, most of which I probably can't even imagine yet. So, I say, let's do it!

Chapter 5

RECIPES

—

Before we dive into the first recipe chapter of the book, I will admit that many of the recipes included in the book are not what most people would consider southern. This is intentional. While I try to include plenty of recipes that any self-respecting southerner would recognize, one ancillary aim of this book is to introduce new foods and flavors to the southern diet. And now for some recipes!

....

SAUERKRAUT

Prep time: 15 minutes | Fermentation time: 2–3 weeks at 20°C | Yield: 1 kg

Ingredients

- 1 kg cabbage (1 medium head)
- 20 g salt (2%)
- You could also add in some caraway seeds (about 0.5% by weight) if you so choose, but these are optional

Instructions

Wash all vegetables well with filtered water.

Remove the core from the cabbage.

Slice the cabbage in thin strips (although this is up to you, depending on your preference)—a food processor can be used for this.

Mix the cabbage with the salt (and caraway seeds if you use them), massaging the salt in vigorously.

Add to vacuum bags, evacuate, and seal.

Ferment away!

Fermentation

This is about as simple as it gets for a lactic acid fermentation. Cabbage, salt, and time are all it takes. Fermentation at room temperature takes 1–3 weeks, depending on your preference for the texture of the cabbage and the amount of lactic tartness. That said, there are some subtle nuances that can make a big difference overall. The fermentation is effectively a 2-stage process, with each stage influenced by the salt content and temperature. The first stage is dominated by *Leuconostoc mesenteroides*, a heterofermentative lactic acid bacteria, meaning that it produces acetic acid, carbon dioxide, and ethanol in addition to lactic acid. These bacteria are particularly salt tolerant and are the first to grow, setting the stage for the second phase. They drop the pH of the solution by producing lactic and acetic acid, thereby inhibiting pathogenic bacteria. Additionally, the carbon dioxide produced creates an anaerobic environment, further limiting aerobic bacteria. The environment created encourages the second wave of bacteria to take over, the *Lactobacillus plantarum*. These are homofermentative bacteria that produce primarily lactic acid, further reducing the pH and creating the signature lactic tartness associated with sauerkraut. A lower salt content (closer to 2%) and lower temperatures (closer to 18°C / 64°F) will favor the heterofermentative *Leuconostoc mesenteroides*, while higher salt content (closer to 4%) and higher fermentation temperatures (closer to 30°C / 86°F) favor the homofermentative *Lactobacillus plantarum*. I encourage you to try both extremes to determine which resulting product you prefer.

Don't believe me that mold is virtually inevitable when fermenting in jars or crocks? This sauerkraut is less than a week old with evident mold growing at the top.

KIMCHI

Prep time: 60–90 minutes (including salting time)
Fermentation time: 5–20 days | Yield: 1.5 kg

Ingredients

- 1 head napa cabbage (about 1 kg) (could substitute with bok choy, although personally I prefer the napa cabbage)
- About 6 cloves garlic
- Thumb-sized fresh ginger, peeled
- 4 g sugar (optional)
- 10 mL fish sauce or shrimp paste
- 5–25 g gochugaru (Korean chili flakes) or an equivalent amount of chili oil (depending on how spicy you like it)
- 250 g daikon radish
- 1 bunch scallions, trimmed
- 2–4% salt by mass (weigh the other ingredients and multiply by 0.02–0.04 to determine the mass of salt to add)

Instructions

Wash the cabbage well with water, making sure to open it up and get any dirt out of the hidden cracks. Cut the base off the cabbage and cut the rest into roughly 2.5 cm squares. Set aside in a bowl large enough to accommodate all the ingredients.

Rinse the scallions well with water and trim the ends off. Cut the scallions into approximately 2.5 cm pieces. Add to the bowl with the cabbage or set aside for use in spice paste.

Rinse the daikon radish well with water. Peeling is optional depending on the condition of the radish. Slice the radish into thin rounds, matchsticks, or grate, depending on your preference. Add to the bowl with the cabbage and scallions.

Add the salt to the vegetables and stir until well combined. Let sit for 30 minutes to 1 hour.

Add the garlic, ginger, sugar, fish sauce or shrimp paste, and chili flakes or chili oil to a blender. You can also add the scallions to the spice paste at this point or leave them as 2.5 cm pieces with the other vegetables. Blend until any large pieces are broken down and the paste is well mixed.

Add the spice paste to the salted vegetables (which should have a decent amount of liquid now from the salting) and mix well until all the vegetables are coated with the spice paste.

Pack the ingredients into a vacuum sealed bag at this point, evacuate, and seal. You can cut corners from the bag to release gas if it expands too much and to test the liquid. Remember to pour some liquid out to test the pH. Never stick a pH strip or probe directly into your ferment.

Fermentation

Keep the fermenting mixture in a cool room temperature (about 18°C / 64°F), dark location for about 5–20 days, depending on your preference. A somewhat cool fermentation is the traditional method, although we have fermented it as high as 35°C (95°F) with perfectly fine results. Like sauerkraut, kimchi similarly undergoes a 2-phase fermentation, the first being heterofermentative, in which acetic acid, carbon dioxide, and ethanol are produced as well as lactic acid, followed by a homofermentative phase,

in which primarily lactic acid is produced. The main microbes involved
are *Leuconostoc mesenteroides*, *Lactobacillus brevis*, and *Lactobacillus
plantarum*. After 1 day, remove a sample of the liquid and test the pH.
It should be below 4.6 at this point. If it is not, check it again after another
day. If it has not dropped below this critical pH level, discard the ferment,
wipe your tears away, pour yourself a fermented beverage, and start again,
ensuring that you follow the procedure carefully. After 5 days (of a successful ferment), remove a sample of the kimchi and taste. If it's to your liking,
move the fermentation to the refrigerator to retard further fermentation
and use it as a condiment in anything from noodle dishes to burgers. If it's
not quite ready yet, continue to test it on a daily basis until the fermentation
is to a point that you prefer.

A bag of kimchi
happily (and
anaerobically)
fermenting away.

CHOW-CHOW

Prep time: 30 minutes | Fermentation time: 1–3 weeks | Yield: 2.5 kg

Ingredients (for a slightly spicy version)

- 1 kg cabbage (1 medium head)
- 270 g cucumbers (1 large or 2 small cucumbers)
- 665 g unripe tomatoes (2 tomatoes, green or light red)
- 210 g red bell pepper (1 pepper)
- 180 g jalapeños (3 peppers)
- 300 g sweet onion (1 onion)
- 28 g turmeric root
- 53 g salt (2%)

Chow-chow ready for its fermentation
journey on day 0.

Instructions

Remove the core from the cabbage.

Remove the stems, pith, and seeds from the peppers.

Peel the onion and remove the ends.

Dice vegetables and turmeric pretty fine (although this is up to you, depending on your preference)—a food processor works great for this.

Mix all ingredients with the salt.

Add to vacuum bags, evacuate, and seal.

Ferment away!

Fermentation

Chow-chow is the ultimate kitchen sink lactic acid ferment. Traditionally, the late season vegetables remaining in the garden were made into chow-chow. So, feel free to use any combination of vegetables that you so desire. Typically, cabbage and cucumbers are components, but there are plenty of sweet onion and green tomato chow-chows as well. Or, just throw them all together like in this one. Turmeric is a common ingredient, too, and adds a nice earthy musk to it. Chow-chows can be spicy, tangy, or sweet. If you

Chow-chow progressing through its fermentation journey.
From left to right: Day 3 showing the expansion from gas production; day 5 after the gas has been expelled and the pH checked; day 7 and just about done on its journey.

want a sweeter chow-chow, you could always add some sugar once it's done fermenting. Just make sure to keep it in the fridge at that point, as it will start refermenting if left out.

As far as the fermentation is concerned, it will be very similar to the kimchi ferment. Room temperature is great for the ferment (anywhere in the 20–25°C / 68–77°F range works well, although don't worry if you go a little above or below this range). I would let this one go a little longer than kimchi, though, especially if you're using cabbage, onions, and peppers, which take at least 1 week, if not 2, to reach suitable textures. However, this is completely up to your preference. If you think it's ready, then eat it. Chow-chow is like a southern version of kimchi. Use it as a condiment with whatever foods you like. I love it on burgers and hot dogs. Enjoy!

PICKLED WATERMELON RINDS

Prep time: 30 minutes | Fermentation time: 3 days to 3 weeks | Yield: 1 kg

Think of these more as a pickle than as anything sweet. The rind contains far less sugar than the core of the melon. Plus, during fermentation the sugars will be consumed, leaving a sour, pickled product rather than something sweet on the palate. As such, I've included more traditional spicy pickling ingredients for our watermelon rinds. Enjoy!

Ingredients

- 1 kg watermelon rinds
- 5 g coriander seeds
- 5 g yellow mustard seeds
- 30 g garlic
- 10 g dried chipotles
- 21 g salt (2%)

Instructions

Process the watermelon to remove the sweet fruit on the inside (some of which you can add to your tepache—see the recipe in chapter 20). Retain the rinds after removing the fruit. Now, remove the outer green skin from the rinds. What remains should be the pale green/white rinds only.

Cut the remaining rinds into thin strips about 2 cm thick (although this is up to you, depending on what you like) and add to a bowl.

Mix with the other ingredients, including the salt.

Add the ingredients to a vacuum bag, evacuate, and seal.

Allow to ferment at room temperature for about 1 week.

Fermentation

This is a straightforward lactic acid ferment. Check the pH after 1 day or 2 to ensure that it has dropped below 4.6. Make sure to release the pressure in the bag and reseal it if it starts to get too full (trust me on this one, or you'll be cleaning up watermelon rinds from all over your kitchen). Taste the pickled rinds after about 3–4 days. Once they're at the desired texture and flavor, they are ready to enjoy. These are extremely versatile too, as the rind itself has very little flavor. You can add any type of spices you would like, and the pickled rinds will absorb the flavor.

I find it very satisfying making these the way that southern grandmas intended. Unfortunately, nowadays you are hard pressed to find any fermented pickle rinds. Rather, you are more likely to find ones that have been quick pickled in vinegar and may even have green food coloring added to them. Why? The rind was never that green to begin with, so why would you want them a bright, kelly green color? It's a very easy and satisfying ferment and a great way to use a product that would otherwise be considered waste. It allows you to access the nutrients from a part of the fruit that is not normally consumed.

Some watermelon rinds on their way to becoming delicious pickled watermelon rinds.

71

PINEAPPLE HABANERO FERMENTED HOT SAUCE

Prep time: 30 minutes | Fermentation time: 2–3 weeks at 20°C | Yield: 900 g

Ingredients

- 530 g pineapple
- 340 g habanero peppers
- 60 g ginger
- 19 g salt (2%)

A soon-to-be pineapple habanero hot sauce.

Instructions

Wash the vegetables well with filtered water.

Remove the top, skin, and core from the pineapple.

Compost the pineapple top, but retain the skin and core for tepache (see the recipe in chapter 20).

Cut the pineapple into 2.5–3 cm pieces.

Remove the stems and cut the habaneros in half.

I recommend wearing gloves when working with hot peppers. If not, wash your hands well before touching your eyes or going to the bathroom. Trust me, I speak from experience on this one.

You can remove the pith and seeds if you would like, but this is time consuming and unnecessary because of the additional processing that will occur.

Cut the ginger into small pieces, roughly 2–3 cm long.

Mix the salt with the other ingredients and add to a vacuum bag.

Evacuate the bag, seal it, and ferment away.

Check the pH of the liquid in the ferment after 1 day or 2 by cutting a corner of the bag and pouring out some liquid.

Reseal the bag and let it continue to ferment, releasing gas every couple of days.

After a couple of weeks, the ferment should be completed. It's hard to test this one before processing, so smell it and sample the liquid instead. If they are to your liking, then call it done.

Blend everything in a blender to break it down as much as possible.

If you simply add the blended hot sauce to bottles at this point, it will be chunky, and the liquid will separate from the solids. Instead, it's a good idea to run it through a food mill. You could also manually sieve it, but a food mill is a great tool that makes this part of the process a lot easier. An all stainless one will only run you $30–40, and it's worth it in my opinion.

Run the blended hot sauce through a food mill using the finest sieve. Now you can add the hot sauce to bottles. Store them in the fridge to maintain the live cultures (plus, they could explode from continued fermentation activity).

Putting the blended ferment through the food mill to homogenize the hot sauce. Don't forget to retain the solids to make chili powder.

Save the solids that were retained by the sieve. Spread them in a thin layer on drying racks and dehydrate them at a low temperature. Once dry, grind them in a spice grinder for some amazingly tasty chili powder.

Fermentation

This is a simple and delicious fermented hot sauce. The pineapple flavor complements the habaneros perfectly, with the ginger adding in a little kick as well. Plus, we can use the core and skin of the pineapple for making tepache (see the recipe in chapter 20), so (almost) nothing is wasted. Fermentation at room temperature works great (20–25°C / 68–77°F). If you go below that a bit, it will just take a little longer. If you go above it by a bit, that's fine too, it will just go pretty quickly. Many people love to let their fermented hot sauces go for months and months, which I don't really understand. The fermentation is pretty complete after a couple of weeks, with all the sugars converted to lactic acid. I like to arrest it before the distinct flavors of the ingredients are lost. But this is your hot sauce, so do with it what you will. If you're going for the long haul and you're fermenting it in a jar or some other container that allows oxygen ingress, be sure to check it every so often for kahm yeast or mold growth.

The solids spread out on a drying rack ready to go into the dehydrator to make delicious chili powder.

PINEAPPLE HABANERO FERMENTED PEPPER JELLY

Prep time: 60 minutes | Fermentation time: 2–3 weeks at 20°C | Yield: 2 L

Ingredients

- Use the same ingredients as for the pineapple habanero fermented hot sauce. This time, though, you'll want to take a little more time processing the vegetables (see below).
- About 1 kg sugar
- Liquid fruit pectin (one 3 oz. packet, or about 90 ml)

Instructions

Deseed the habaneros and remove the pith as well (please wear gloves for this).

Grate the ginger.

Cut the pineapples and habaneros into small pea sized pieces. Alternatively, you could use a food processor for this step. Just don't over process the veggies. You want to maintain some recognizable chunks of peppers and pineapple for the jelly.

Otherwise, follow the same instructions as for the hot sauce.

Ferment away!

Now, instead of blending and milling, we're going to turn our fermented product into a pepper jelly.

Pour the fermented mixture into a pot, liquid and all. Add the sugar and slowly bring to a boil while stirring fairly often so that the sugar doesn't burn on the bottom. Once boiling, turn it down a bit and boil for 10 minutes.

Stir in the pectin and boil for an additional minute. Remove it from the heat and let it cool for a few minutes.

Add to jars, cap, and let cool to room temperature. Move to the refrigerator once cooled to room temperature.

If the jelly doesn't set up properly, pour it back into a pot, bring to a boil again, add another packet of pectin, boil for another minute, and remove from heat. Follow the same instructions as before. If it still doesn't set up, enjoy it as a spicy, sweet topping for ice cream or pie.

....

OGI

Prep time: 30 minutes | Fermentation time: 3–5 days | Yield: 300 g

This is a traditional west African fermented porridge. It can be made with corn, millet, sorghum, or probably many other grains. The recipe below is a version adapted from the Yoruba of Nigeria, one of the major ethnic groups, or tribes, of that country. It was brought to me by my friend and colleague Folarin Oguntoyinbo. It uses corn, which is also very southern, so I thought it would be a perfect recipe for this book. And it's similar to sour corn, the easy lactic acid fermented corn that has been enjoyed by southerners for as long as anyone can remember, just essentially taken one step further. Enjoy!

Ingredients

- 454 g dried field corn (not sweet corn)
- Filtered water

Instructions

Place the corn in a large pot or jar and add filtered water to twice the height of the corn.

Allow the corn to soak for 24–48 hours.

Drain the water and blend the corn.

Use a food mill with a fine sieve to mill the corn mush.

Rinse the recovered corn starch multiple times with water to remove any remaining detritus. The only remaining material should be the fine white corn starch.

Soak the remaining corn starch in filtered water (enough to cover the starch to twice its height), covered with a cloth or coffee filter, for 2–3 days or until the mixture begins to bubble and smell sour.

Once soured, the water can be drained and the starch can be stored in a jar for a couple of months in the refrigerator.

To cook the ogi, take a portion of the starch and heat 4–5 times as much filtered water in a pot to boiling. Once boiling, slowly stir in the starch and continue stirring until the water is incorporated and the mixture develops a gelatinous porridge consistency. Add salt, fruit, sweetener, or whatever you fancy, and enjoy.

Fermentation

The fermentation is the result of lactic acid bacteria on the grain. You don't need to add a culture, but you can do so to ensure a good ferment. The main lactic acid bacteria that are involved in the ferment are *Pediococcus pentosaceus*, *Lactobacillus brevis*, and *Lactobacillus plantarum* (Teniola and Odunfa 2002). Yeast may get involved as well, but generally only when the ogi begins to spoil; this happens if it's allowed to ferment for overly long. Spoilage will be evident from a hydrogen sulfide smell (rotten eggs or, well, flatulence, if you're not aware). Some online recipes recommend adding a culture from kombucha or kefir, but I would strongly advise against this, as they contain yeast that may hasten the spoilage. If you're going to add cultures, try to ensure they are only the above-mentioned lactic acid bacteria.

....

COWPEA NATTO

Prep time: 30 minutes | Fermentation time: 24–48 hours | Yield: 450 g

Ingredients

- 450 g cowpeas (lady cream peas work really well)
- 1 g natto culture (*Bacillus subtilis*)
- Sterilized water

Instructions

Wash the beans to remove any dust and debris.

Soak the beans overnight in enough filtered water to cover the beans by at least 15 cm of water.

Drain the beans and add them to a pot with plenty of filtered water to cover the beans.

Boil, uncovered, for about 30 min (they should be al dente—done, but not mushy, with a bit of a snap still).

Drain the beans in a sanitized colander, but do not rinse or shock them with cold water (you want them pretty warm to start the ferment).

Place drained beans in a sanitized, nonreactive pan or dish with sides that are high enough to be above the level of the beans.

Activate your natto culture by stirring it into 50 g of boiled and cooled (but still warm, about 38–49°C / 100–120°F) water. This is your starter culture.

Pour the starter over the warm beans and stir using a sanitized utensil.

Spread the beans in a 3 cm layer in your pan or dish.

Cover the pan or dish tightly with aluminum foil and poke holes with a toothpick every 3 cm.

Ferment the natto at 38°C (100°F) for about 24 hours.

You'll know it's done when the beans are covered in a white film that forms sticky, gooey strings when stirred. Oh yeah, and you'll smell it too.

Once done fermenting, cool the natto and store it in a covered container in the fridge (it should keep for about 1 month, possibly longer).

Bonus Round: try drying the natto in a dehydrator, grinding it up, and using it as a fish sauce/umami bomb in dishes. Or, don't grind it and snack on the dried beans as part of your trail mix. Very tasty!

Fermentation

So, about that smell. This is an alkaline ferment, as opposed to most others that are acidic. The *Bacillus subtilis* that is used to produce natto (and other similar ferments in Africa) metabolize the proteins in the beans, producing ammonia. So, not only will it smell very funky and fishy, the ammonia also raises the pH above 7 into the alkaline range. Because of this, sanitation is critical for this ferment. You don't have the added protection of being below that magic pH value of 4.6. And I'm not kidding about the smell. You'll know when it's getting close to being done. A while back, we were making natto in our facility, which we share (in a completely separate part of the building, mind you) with the transfer services office. We started the fermentation the afternoon prior. I went into work that morning and didn't notice much of a smell. I ran out to do some errands and got a call from someone in transfer services. They smelled something weird and wanted to know if it was coming from us before they called the fire department to check for gas leaks. I told them that I didn't think it was coming from us, as I hadn't smelled anything untoward when I was in that morning. And then I returned to the facility. As soon as I walked in, the funk hit me. Yup, the natto was getting ripe. I called transfer services back and informed them that we were, in fact, the malodorous malefactors.

I won't even say that natto is an acquired taste. I think it's more a love it or hate it situation. I really like the funkiness and extreme umami character, almost like an overripe bloomy rind cheese or fish sauce. If you're not brave enough to eat it on its own, try adding it to soups, stews, rice, or other dishes for flavoring. You'll lose the probiotics by cooking it, but the flavor it adds to dishes is quite savory. Hence the suggestion to dehydrate and grind it as an umami rich flavor booster. I love the flavor of cowpeas, and I think they make a decent substitute for the standard natto soybeans that are typically used. And cowpeas are just so darn southern. From our trials, the lady cream peas are the best of the cowpea bunch for natto. In general, cowpeas won't get quite as funky as the natto soybeans, and this may also be an added bonus if the standard natto is a little much for your taste.

....

BARLEY KOJI

Prep time: 60–90 minutes | Fermentation time: 46–48 hours | Yield: 1 kg

Remember, koji technically refers to *Aspergillus* growing on a substrate. Traditionally, the most common substrate for growing the mold has been rice. However, I'm providing a recipe for barley koji because this is the substrate with which I'm most familiar. If you'd like to learn more about our research into this area (or if you need a good sleep aid prior to bed), you can read our riveting account, "Characterization of Unmalted Barley Treated with *Aspergillus oryzae*" (Williams, Parker, and Taubman 2021).

Ingredients

- 1 kg pearled barley
- 0.8 g red rice koji spores (feel free to use other *Aspergillus* strains, though)
- 29 g barley, wheat, or rice flour
- Filtered water

Instructions

We begin here by using the red rice koji to make the koji starter. This is strongly recommended so that you don't have to deal with the very fine mold spores that otherwise tend to get everywhere. First, sterilize 29 g of flour. This can be done by lightly toasting the flour in a pan on medium heat until it browns slightly (don't over toast or burn it). You just want to make sure that any microbes on the flour are not viable. Once the flour is cool, mix in the red rice koji spores until well combined. This starter can be stored indefinitely at room temperature in a covered container.

Rinse the pearled barley well with filtered water to wash off any dust and debris.

Soak the pearled barley in filtered water for 24 hours. It's best to keep the soaking barley refrigerated, if possible, to mitigate any risk of spoilage.

After soaking, rinse the barley with filtered water and let it drain for 1 hour.

Steam the barley for about 20 minutes. It should be done, but al dente, not over cooked and mushy.

Transfer the steamed barley to a nonreactive baking tray (perforated is best if you have it for maximal air flow, as this is an aerobic process) lined with a sanitized tea towel.

Stir the barley until cooled to 25°C (77°F).

Inoculate the barley with 5.25 g of the prepared starter by stirring it thoroughly into the steamed barley. It's probably easiest to do this with your hands (wear nitrile gloves so as not to introduce any additional microbes to the process). The barley should be evenly distributed in the pan at a depth of no greater than about 3 cm.

Incubate the barley in an environmentally controlled chamber at 35°C (95°F) and 90% relative humidity. A converted refrigerator or freezer works great for this. See the "Tools of the Trade" section in the introduction for details on how to construct one.

After 24 hours, stir the barley (again, gloved hands work best). You probably won't see much growth at this point, but you should shortly after. Make a few lengthwise furrows in the stirred grain after evening it out in the pan. They should look like furrows in a field. This creates a greater surface area for oxygen exposure, and it also allows for more heat to escape during fermentation. If the heat builds up too much it can cause the fungi to sporulate too soon.

After another 22 hours, the barley should be fully coated in a nice, off-white mold. At this point, it is done.

To prevent further growth and sporulation, place the barley koji in a sealed container and refrigerate for 4 days.

After 4 days, dry the barley koji in another baking tray in the chamber at 35°C (95°F) and 15% relative humidity until a 35% moisture loss has occurred. To determine when this is, weigh the barley prior to drying. Calculate a 35% reduction in that mass. Once the dried barley has reached that 35% reduction in mass, it's dried and stable.

Now the barley koji can be stored at room temperature until you're ready to use it.

Fermentation

We used the red rice koji spores (purchased from www.fermentationculture
.eu, which is a great source for koji spores in general), which are a strain of
Aspergillus oryzae, for the study noted above. We used this strain primarily
because the flavor profile suited our needs for the study we were conduct-
ing. However, we have used many different strains of *Aspergillus oryzae* as
well as a strain of *Aspergillus luchuensis* in our lab. Feel free to experiment
and choose one or a few that you like for your needs. They all produce fairly
unique flavor profiles. The *A. luchuensis* is very interesting, as it produces
a lot of organic acids, especially citric acid, making a fairly tart and citrusy
koji barley.

As *Aspergillus* grows, the mycelium work their way into the substrate,
in this case the barley, and excrete enzymes to digest the material. In general, this is how filamentous fungi obtain necessary nutrients. Whereas we
put food in our body and then digest it, filamentous fungi excrete enzymes
into their food to digest it external to their bodies (in so much as filamentous fungi have bodies), and they then take up the resulting nutrient molecules. In this case, the *Aspergillus* excretes starch- and protein-degrading
enzymes to degrade them into sugar molecules and amino acids, respectively. These molecules are then taken up by the fungus and used for fuel
and growth. In our study, we were interested more in the action of the
starch-degrading enzymes than the protein-degrading enzymes. As such,
we used a fairly high temperature for incubation (35°C / 95°F), which leads
to greater activity of the starch degrading enzymes. If you are more interested in umami character from amino acids, the incubation temperature
could be reduced to closer to 30°C (86°F). This will preferentially activate
the protein-degrading enzymes instead, resulting in more amino acids and
greater umami character.

SHIO KOJI

Prep time: 10 minutes | Fermentation time: 10–14 days | Yield: 600 g

Ingredients

- 250 g dried barley koji
- 50 g salt (anywhere from 10–30% of the mass of the koji)
- 350 g filtered water

Instructions

In a mixing bowl, break up the barley koji until it is mostly individual barley kernels.

Add the salt and mix this in well with the koji.

Stir in the water and mix until combined.

Add the mixture to a jar and place a clean towel or coffee filter over the top and secure it with a rubber band. Store the jar at room temperature away from direct sunlight.

For the next 10–14 days, open the jar once a day and stir the mixture. Replace the towel or coffee filter until the next day.

It is done when the mixture smells a little yeasty and sour and is the consistency of a porridge. The warmer it is, the faster the ferment will go.

At this point, put a lid on the jar and store in the refrigerator for up to 6 months. The more salt that is used, the longer it will keep.

Fermentation

Shio koji (which translates from Japanese to salt koji) is effectively the nascent stages of miso, as in baby miso. The salt largely inactivates the *Aspergillus*, but the enzymes that were produced when making the barley koji are still viable and in solution. Lactic acid bacteria and possibly some yeasts feed on the sugars produced from the *Aspergillus* enzymes. Those are the microbes that produce the activity observed during the production of the shio koji. As such, it is an active ferment with additional enzymes from the *Aspergillus* that can be applied to a number of substrates, like meat, and used as a marinade. It has a number of other possible applications, but the protein-degrading enzymes in the shio koji make it a quick and super effective marinade that can be applied to any type of meat. It's not just the enzymes, though. The high salt content draws moisture out of the meat, just like when plain salt is added to the meat. Additionally, the flavors from the organic acids produced by the other microbes in the shio koji add an incredible flavor-boost to the meat.

To use it as a marinade, just rub it all over the meat you're marinating. You can sieve it or even blend it prior to applying it if you want a smoother paste for application. Leave the meat at room temperature on a rack with drainage underneath for the water that is drawn from the meat. For smaller and less tough cuts of meat (or fish), 1 hour is likely sufficient. For larger or tougher cuts of meat (or whole birds), a few hours may be necessary. You can apply it and also leave the meat in the refrigerator, but this will seriously hamper the activity of the shio koji, so you may need a couple of days in this case. I would advise shooting for a shorter marination time to begin with and increase the time incrementally until you've found the right time range for the particular cut of meat. Better to underdo than to overdo it, as you can easily end up with an overly dry and salty piece of meat if you're not careful. After marinating, scrape the shio koji off and pat dry with a paper towel. You're now ready to cook and enjoy your tender, succulent piece of meat!

MISOS

Prep time: 30 minutes | Fermentation time: 3 months to 2 years

The barley koji or koji in general has so many applications, too many in fact to include in this book, but this is one that you should definitely add to your quiver of go-to fermentations. It takes a while, but it's sure worth it, and it has plenty of applications of its own. Plus, there is an endless variety of options in the types of misos that you can make with it. These are just a few ideas from what we have done in our lab.

Ingredients

I'm leaving these instructions fairly nonspecific so that you can tailor them however you like.

For dark or salty miso:

- Protein/carbohydrate component of your choosing
- Barley koji in an amount that is ½ the mass of the protein/carbohydrate of your choosing
- 10% (of the combined mass of the protein/carbohydrate and koji) salt

For light or sweet miso:

- Protein/carbohydrate component of your choosing
- Barley koji mass equivalent to that of the protein/carbohydrate
- 5% (of the combined mass of the protein/carbohydrate and koji) salt

Some of the protein/carbohydrate with koji combinations we have used with success are given in table 5.1.

TABLE 5.1. Some of the misos produced (both dark and light) using the following base ingredients (as the protein/carbohydrate) and different barley kojis

Protein/carbohydrate	*Aspergillus* spores used to make barley koji for miso
Mushrooms (we used a combination of chestnut and pink oyster, but use whatever species you like)	Red rice koji (*Aspergillus oryzae*)
Potatoes and parsnips	*Aspergillus luchuensis*
Sweet potato (purple and garnet)	Red rice koji, white rice koji (*Aspergillus oryzae*)*
Nixtamalized corn	Red rice koji
Winter squash	Red rice koji
Beets	White rice koji
Beans (we've used soy, cowpeas, limas, garbanzos, and several others)	Red rice koji

*Red rice koji and white rice koji are different strains of the same species of *Aspergillus*.

Instructions

For the beans, soak them overnight in filtered water and then boil them until they are softened a bit, but not quite cooked through. You just want them degraded sufficiently so that they are amenable to enzymatic activity.

Wash all the raw vegetables well in filtered water.

Dice the vegetables finely. You can use a food processor for this step as well.

Grind the barley koji to a flour.

Combine the protein/carbohydrate with the koji and salt and mix the ingredients well. You can add the mixture to a food processor as well, but you don't have to. If not, mash them up pretty well as you mix them together. Everything will break down over time, but the more broken down the elements are in the beginning, the faster you will end up with a smooth paste-like consistency.

Add the mixture to vacuum bags, evacuate, and seal.

Light misos will be done in 3–6 months and dark misos can go for 1–2 years or more.

Fermentation

By fermenting the miso in vacuum-sealed bags, you are virtually eliminating the threat of aerobic bacteria spoiling your miso party. The *Aspergillus* is inactivated by the salt, but its enzymes are already in the mix, breaking down the protein/carbohydrate substrate. This produces rich umami character from the resulting amino acids as well as sugars to feed the lactic acid bacteria that you're trying to encourage. The more salt you add, the more this selects for only the most salt-tolerant bacteria and the longer the overall process takes. However, more salt also means the resulting product is more stable in the long term. So, you may be sacrificing some flavor production by retarding some bacteria that would normally be contributing to the flavor profile, but you make up for it with a product that can last for years. And, believe me, I've never had one of our misos and thought, geez, I really wish this had more flavor. They are generally salty, rich, umami bombs. In fact, the sweet potato misos might be some of the best things I have ever put in my mouth.

A tale of two misos. The soy miso on the left is about 6 months old; the miso on the right is more fully fermented at 2 years old. Note the difference in colors and textures.

Since the process really creeps along, it is rare that you will even have to burp the bags to release the gas. Just keep them in a dark, warmish (25–30°C / 77–86°F) place and check on them every so often. You probably won't notice much activity other than the gradual breakdown of the material into more of a paste-like consistency and the production of the resulting liquid (which is an umami-rich tamari that you can use separately!). You can keep the ferments cooler than this as well. The fermentation will just be a little slower overall.

LIMA BEAN TEMPEH

Prep time: 45 minutes | Fermentation time: 18–24 hours | Yield: 454 g

Ingredients

- 454 g dried lima beans
- 27 g white vinegar (2 tbsp) or any other lightly flavored vinegar
- 3 g starter culture (*Rhizopus oligosporus*)
- Filtered water

Instructions

Soak the lima beans overnight, covered by at least 15 cm of filtered water to account for their increase in size.

Drain the beans and add them to a pot with plenty of filtered water to cover the beans.

Boil the beans for about 15 minutes. You want them just cooked through, but not overcooked. If they start to split, you've gone too long. Soaked limas really don't take very long at all, so start checking them after about 10 minutes.

Drain the water from the beans, but don't cool them.

Using gloves (since they'll still be warm!), de-hull the beans. Limas are about the easiest beans to de-hull. I like to lightly pinch the bean at the midway point (lengthwise) on the convex side and give a little squeeze. This should squirt the bean out, leaving the hull in your pinched fingers.

Dry the beans well using either a clean towel or a hair dryer.

When the beans are lukewarm (35–38°C / 95–100°F) mix in the vinegar using gloved hands.

Add the starter culture and mix well with gloved hands.

At this point, you have a couple of options. You can add the beans to a sanitized, non-reactive open pan or dish in an even layer about 4 cm deep or put them in sanitized Ziplock sandwich bags. If you use bags, first poke holes using a sanitized needle or other thin, sharp, pointy object every 2.5 cm or so on both sides of the bags. Now pack the beans in them so they are evenly distributed to a depth of about 4 cm. Squeeze out the air and seal the bags so the beans are tightly held together and evenly distributed in the bags.

Ferment the tempeh at 29–33°C (85–91°F), although I advise erring on the lower side of this range. The fermentation will generate heat. If the ferment gets too warm, the mold can sporulate or even die. To monitor the temperature, place a thermometer in the beans or as close as possible to them so that you are maintaining the correct temperature in the ferment itself, not just in the surroundings.

This is a quick one, so start checking your tempeh after about 18 hours. In general, it should not take longer than 24 hours. Once you notice the white mycelium coating the beans and knitting them together, turn the heat off, but leave the tempeh where it is for another 6–12 hours. It is ready when the beans are totally coated and held together in a nice, firm cake that holds together on its own.

Adding the inoculated lima beans to bags with holes poked in them.

The bagged, inoculated lima beans ready to go into the fermentation cabinet.

Fermentation

For this tempeh, be sure to use the *Rhizopus oligosporus* culture rather than the *Rhizopus oryzae*. *R. oligosporus* does better with non–soybean tempehs, whereas the *R. oryzae* is generally the better option for traditional soy-based tempehs. If you purchase your cultures from Cultures for Health (https://culturesforhealth.com/), which is a great option for cultures in general, the *R. oligosporus* culture is even labeled as "Soy-Free Tempeh," while the *R. oryzae* cultures are labeled simply "Tempeh." This is a very straightforward ferment and about as short as they come; however, there are some real pitfalls to be aware of. First, do not overcook the beans. If you do, you'll end up with a mushy mess that doesn't knit together very well into a nice firm cake. Second, make sure to thoroughly dry the beans. If you do not, you may end up drowning the spores and get no fermentation at all. Third, make sure to monitor the temperature of the ferment itself rather than just the surroundings. It can really crank out some heat that, best case scenario, can cause the mold to sporulate too quickly (you'll notice greenish black areas on your nice, white mold) or even die. If the tempeh sporulates, it's not the end of the world. This is actually commonly consumed in Indonesia and referred to as *tempe bosok*, or rotten fermented soybean. That latter name should be a clue, though, as to the flavor of this sporulated tempeh. It is ammoniated and pretty funky, even borderline rancid. It's edible, but a decidedly acquired taste.

Tempeh production is an alkaline ferment. This means that the pH is above 7, which leaves it susceptible to pathogenic bacteria. So, do not consume tempeh raw! You must cook it first. Yes, you are killing the microbes, but this may be a good thing, as some could be harmful. Even so, tempeh is one of the most healthful foods you can consume. It is high in protein and fiber and also has antihypertensive, antidiabetic, antioxidative, and antitumor properties, as described earlier. Plus, it's not overly flavorful when raw. Rather, it has a light, nutty flavor that is great at absorbing any flavors that are added to it when cooking. I love it in a number of Asian dishes but also in simple dishes like a vegan alternative to a Reuben. Now, go make yourself some tempeh, cook it up, and enjoy!

Part II

MEAT

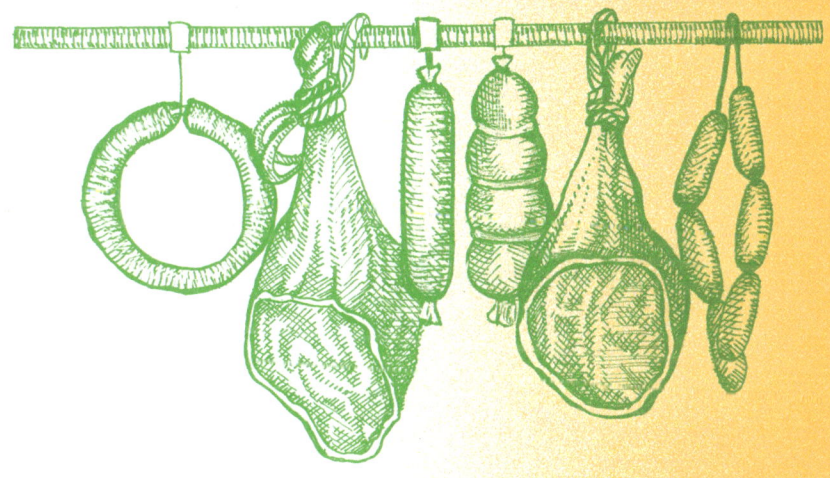

Chapter 6

PAST

—

Pork was the predominant livestock in the southern United States, and preserving large cuts of meat using fermentation was a practical and efficient approach. This is where we got country ham. True country ham is not just salt cured but also fermented by lactic acid bacteria over a long period of time. Other fermented products that made use of less choice cuts led to various farmers' sausages or country-style sausages such as Tom Thumb/Dan Doodle. Let's look more closely at some of these products.

COUNTRY HAM

"No self-respecting Southern pig can imagine a higher distinction than becoming, in due course, a Virginia ham—spicy as a woman's tongue, sweet as her kiss, as tender as her love" (Fishwick 1964).

Smithfield, Virginia, is the birthplace of country/Virginia ham. European techniques, brought by colonialism on tidewater Virginia in the seventeenth and eighteenth centuries, combined with local Indigenous expertise to produce the Virginia ham. It is a result of the creolization of the colonists, where European ideals met colonial realities and Indigenous and African traditions. Meat preservation practices in early colonial Chesapeake derived from both English and Algonquin traditions. The influence from the enslaved peoples of West African descent was minimal, as this was still the earliest days of colonization in the reprehensible times of English indentured servants and prior to the abysmal shift to enslaving African people in the colonies. By the time of this shift, in the mid seventeenth century, meat curing practices in the Chesapeake were already established. Algonquins were already smoking meats to preserve them and possibly drying them as well. The tradition of salting/curing (with salt, sugar, and maybe saltpeter or hickory ash) meat for preservation was English. Smoking meat was initially an Algonquin practice, with later influence from Afro-Caribbean techniques. Thus it came to be that hams were cured and then hung in a smokehouse over low, slow hickory fires for up to 6 weeks. Traditional English meat preservation methods such as pickling did not work in the heat and humidity of colonial Virginia. Dry curing was much more effective at preventing the spoiling of meats in the harsh and variable weather conditions of the Chesapeake.

Pork, the livestock of choice in the early colonies, is particularly high in saturated fat, even relative to other livestock meats. And saturated fats have a slower oxidation rate than unsaturated fats, which is bad if you're trying to lose weight (the fat takes longer to burn for energy and is, therefore, stored as adipose tissue—fat) (DiNicolantonio and O'Keefe 2017) but great if you're a seventeenth-century colonist relying on cured meats to not spoil and to keep you alive and all that (Edwards 2011).

The hardwood forests of colonial Virginia and the Carolinas provided not only ample wood for smoking meats but also plenty of acorns and nuts for making fat, delicious pigs. It didn't take long for those ever-observant English settlers to notice the relationship between the quality of an animal's diet and the quality of its resulting flesh. John Lawson, who helped found the colony of North Carolina, noted in his seminal 1709 travel guide for English settlers, *A New Voyage to Carolina*, how tasty the Carolina pigs were after a healthy diet of woodland acorns and nuts; by no means inferior to their English cousins (Lawson 1967).

Curing practices at the time were largely dictated by the availability of raw materials. Salt production was mainly controlled by mother England, in Liverpool. Attempts at American salt production were routinely undercut by English imports, eventually leading to the shuttering of these tech start-ups (Kurlansky 2002). As a result of these global supply chain issues, colonists turned to saltpeter, a common name for potassium and sodium nitrate salts, also a primary ingredient in gun powder, which contributed to its ready availability in the colonies. Saltpeter, which is still used for curing, inhibits foodborne pathogens, namely *Clostridium botulinum, Clostridium perfringens*, and *Listeria monocytogenes* as well as *Salmonella* and *Staphylococcus* bacteria (Lee et al. 2018). Interestingly (if you're a chemist), it's not the nitrate that does the work. The nitrate is reduced to nitrite by bacteria in the meat, which then provides the protection against pathogenic and food spoilage bacteria. Nitrite also adds a lovely red color to cured meats through a complexation with the myoglobin in the blood (Toldra et al. 2015). Alas, then those pesky colonists had to rise up against mother England in the American Revolution and use up all the saltpeter shooting Redcoats, a decidedly less delicious option than using it to make country ham. Sugar, the other common ingredient in curing mixes, is used to balance the saltiness in the cure and potentially add some sweetness; but it also provides an important source of carbohydrates for the flavor-enhancing good bacteria. Unfortunately, it was also generally in short supply in the colonies, despite the fact that by the mid seventeenth century, sugar production was starting to take off in the West Indies (Mintz 1986). The bulk of the exports, however, were sent east across the Atlantic rather than the shorter trip north to the colonies.

As a result of the inconsistencies in the supply of typical curing ingredients, colonists were forced to adopt techniques borrowed from Algonquin and Afro-Caribbean traditions, namely smoking meats for preservation. The smoke draws moisture out of the meat, similar to salting, thereby creating an unsuitable environment for spoilage and pathogenic bacteria. Additionally, the smoke itself coats the surface of the meat with antimicrobial and antioxidant compounds that protect the meat from spoilage microbes and prevent oxidation of the fat that would otherwise lead to a rancid product. Simply put, country ham exemplifies the forced multicultural influences on food preparation and preservation in the colonial southeastern United States. Whereas social and cultural barriers among settlers and Indigenous and enslaved peoples were rarely otherwise breached, they had to be breached for survival purposes when it came to the available foods in the colonies.

TOM THUMB SAUSAGE

In deference to my publisher, UNC Press, and the fact that I've lived in North Carolina for almost two decades, we'll call this sausage Tom Thumb. However, if you cross the border to the north into Virginia, the same sausage is referred to as Dan Doodle. Are the cartoonish names a weak attempt at a euphemism to mask the fact that the sausage is stuffed into a pig's appendix, which may be slightly less than appetizing to those with more delicate palates? Not to mention the fact that pigs don't even have appendixes. Rather, the casing used is actually the cecum, which is the knobby, reticulated near end of the large intestine (the far end being the, you know), also referred to as the hog middle cap, as it's the end-cap of the hog middle (large intestines). Our appendix is like a little inflatable balloon hanging off the cecum, which is probably the reason for the misnomer in this case. Anyway, it's a very large-volume casing, and with the reticulations and knobs, the sausage looks like some sort of alien larva.

Some old-timers claim that the name "Tom Thumb" derives from the fact that the sausage looks like a thumb. I know the farming life can be hard on hands, especially for old-timey farmers, but if your thumb looks like a Tom Thumb sausage, you should probably see a doctor. Obviously, there are references to the fictional character Tom Thumb with respect to the sausage name, but I'm having trouble seeing the connection unless it has something to do with the fact that Tom was constantly being swallowed by one animal or another on his many adventures (Mariani 2014). Tom definitely ended up in a cecum or two along the way, as he generally exited through the back door when this occurred.

As for Dan Doodle, who knows. Here's my weak attempt at a guess, though: Willie Dixon, blues musician, producer, and prolific songwriter, wrote the lyrics and music to "Wang Dang Doodle" in the 1950s, which was recorded by Howlin' Wolf in 1960. Apparently, a Wang Dang Doodle is when a person, especially one from the South, comes into the big city to party and have a good ole time. What does that have to do with the sausage you ask? Dan Doodles (or Tom Thumbs) are a classic southern sausage, which were traditionally enjoyed on New Year's. Pigs were slaughtered in the fall into winter, and the sausage that was made at the time was cured, smoked, and aged for a time, depending on what type of sausage it was. The Dan Doodle, a semidry fermented sausage took anywhere from a few weeks to a few months to reach earthy perfection, just in time for a New Year's celebration. So, it's a southern sausage that sounds surprisingly similar to Dang Doodle and that was enjoyed on a night renowned for celebrations. Sounds like a southern party to me.

There is no definitive recipe, or even process, for making the sausage. However, it is typically spicy, with plenty of sage added (almost like a spicy breakfast sausage), and it's smoked to varying degrees. Most recipes available these days call for using the pork shoulder, the go-to cut for making pork sausage, which has a good meat-to-fat ratio. I have a hard time believing that a prime cut like the shoulder would be used traditionally for making sausage, though. Likely, it was whatever meat was left over after removing the prime cuts, blended with back or belly fat to get the right ratio. In this case, anywhere from 50% meat with 50% fat to 70% meat with 30% fat could be used to make a Tom Thumb. Sometimes even chitterlings (chitlins') or organ meat would be mixed in. Once the meat and fat were ground, mixed with spices and curing salts, and stuffed into the casing, they were likely hung to air dry for a bit and then lightly smoked. Oak or hickory were probably most common, but other hardwoods may have been used, like peach or pecan, depending on availability.

The Tom Thumbs were smoked from a few hours to a few days and then left to hang in the smokehouse sans smoke, where it was cool and slightly humid, perfect for curing, drying, and slowly fermenting. This would last for weeks or months, likely depending on when the pigs were slaughtered relative to the New Year celebration. As a result, the flavor profile must have changed dramatically depending on the amount of fermentation and aging allowed. Some may technically have been a dry sausage, fit for consumption without cooking. That said, apparently for some, the earthiness and saltiness is a bit overwhelming if one attempts to eat a Tom Thumb without boiling it first. That's right, after rinsing the alien larva, they're typically thrown into a pot with some veggies and water and boiled for 1 hour or 2. This leaches some of the salt and earthy character out of the sausage and into a delicious broth to cook your greens in afterward. Then the sausage is sliced and pan-fried to a crispy deliciousness that is served on top of said greens for a New Year's celebration. Which is delicious; however, I can confidently report that fully fermented Tom Thumb is also delicious without any cooking. It takes a while to fully cure and dry the mass of meat in such a large casing, but it's worth the wait. Smoking helps to dry it faster than the unsmoked mold-encased versions, but those are excellent too. But if eaten without cooking it, what will you cook your greens in, you ask? Well, I would suggest a broth made from guanciale, cured pork jowls, also called cheek bacon. Like I tell my students, for every problem, there is a solution. You just have to find it.

I recognize that many of you will probably not have the opportunity to ever try this sausage. It's becoming harder to find these commercially available, as fewer and fewer people even know what they are. You'd be missing out on a literal slice of southern culinary history, though. That is why I wanted to include a recipe. For those of you adventurous enough to make this at home (and with the requisite equipment and supplies for long-term meat fermentations), the recipe is below. I also wanted to include a recipe as a means of preserving a small piece of southern heritage that is rapidly disappearing.

BACKGROUND

—

Traditionally, as is the case with all fermented foods, meat was initially fermented and cured as a means of preservation prior to refrigeration technology. Some of the earliest evidence of drying and smoking meats as a means of preservation comes from ancient Greece and Rome (Zeuthen 2015). This is probably where we get the names "sausage" and "salami" as well, from the Latin words "salsiccia" and "salumen," respectively, which both have the root "sal," or salt, in Latin (Toldra 2012). Whereas meats are still fermented and cured as a means of preservation, they are primarily prepared and consumed these days for gustatory reasons, as in, they're delicious. Even so, it is critical, even more so, arguably, than with other fermented food products, that the meat is preserved properly. To do so, a series of factors, known as "hurdle effects," must be met. These include the following:

- adding nitrate, salt, and/or sugar,

- reducing the redox potential,

- introducing lactic acid bacteria,

- lowering the pH,

- decreasing the water activity (a_w),

- and smoking (Leistner 1992).

Let's look more closely at each of these in turn.

ADDING NITRATE

Nitrite or nitrate, as salts of sodium or potassium, are added to meats prior to fermentation. When making products that are quick cured and meant to be eaten in a short time, sodium or potassium nitrite is used. By contrast, when making products that are fully cured over an extended period of time, such as salami, pepperoni, or cured ham, sodium or potassium nitrate is added along with the nitrite. These are generally added using either cure #1 (also called instacure #1, pink salt, or Prague powder #1, composed of 6.25% sodium nitrite and 93.75% sodium chloride—table salt—and added at 0.25%) for quick cures; or cure #2 (also called instacure #2 or Prague powder #2, and containing 6.25% sodium nitrite, 1% sodium nitrate, and 92.75% sodium chloride—also added at 0.25%) for fully cured products. For either application, it's the nitrite that's doing the work. Nitrite has antimicrobial properties that inhibit the growth of pathogenic bacteria, such as *Listeria monocytogenes*, and *Clostridium*, *Salmonella*, and *Staphylococcus* bacteria by interfering with their ability to produce energy, which eventually leads to their death (Lee et al. 2018). You know that pretty pink color of cured meats? Yup, that's the nitrite too, which forms the color by complexing with the myoglobin in the blood (Toldra 2012). Without nitrite, cured meats would turn a dull gray-brown color, which would probably be quite unappetizing to many people. Nitrite also prevents the oxidation of fats in cured products that can lead to rancidity and off-flavors. Finally, nitrite enhances the flavor of cured meats by contributing to the development of those characteristic cured meat flavors, including an intense umami taste. So, if nitrite is doing all the work, why would you want to add nitrate? Ah, great question. The nitrate in cure #2 is actually reduced to nitrite through the action of the bacteria in the meat. This happens over time, though, which is why cure #2 is used for the long curing process of fully fermented meat products. As the nitrite is slowly produced by the bacteria, it continues to provide protection against pathogenic bacteria throughout the curing process.

However, it's important to note that excessive intake of nitrite can be harmful to human health. Nitrite can react with compounds in the stomach to form nitrosamines, which are potential carcinogens (Karwowska and Kononiuk 2020). Therefore, the use of nitrite in meat preservation is strictly regulated by food safety authorities to ensure that it's used at safe and appropriate levels. This brings up one of my pet peeves—"uncured" meats. Any normally cured meat product that is labeled as "uncured" is absolutely, 100% cured. At least you better hope it is if you're consuming it. Uncured bacon or sausages would be an off-putting gray-brown color, not to mention potentially harmful or deadly, and they would taste awful. Instead of using measured and regulated amounts of sodium nitrite, however, "uncured" products, also sometimes labeled with the seemingly health-conscious statements "no synthetic preservatives" or "no nitrates or nitrites added," use celery juice powder (also called celery powder). Celery happens to have some of the highest nitrate concentrations of any vegetable (many of which have very high nitrate concentrations), taken up from the nitrates in the soil (nitrates being the most common form of nitrogen addition in fertilizers). For commercial applications, the concentrated celery product is treated with a bacterial culture that converts the nitrate in the celery to nitrite. Otherwise, the powder can be left untreated so that the bacteria in the meat do the same thing. Not only can conventionally grown celery, which is notoriously efficient at absorbing pesticides, be used to make the powder for use in organic meat products, but by using the powder, you are still adding nitrates or nitrites to your meat. Nitrate is nitrate, and nitrite is nitrite, regardless of the source. Just because something is "naturally derived" does not mean that it is inherently better. In fact, by using the celery juice powder to cure meat products, the amounts of nitrites added are far less accurate than when using pure sodium nitrite, and the potential for harm is far greater. When I received a marketing email from my favorite local natural foods store, Earth Fare, stating that sodium nitrite was "Booted Ingredient #109," and including the following statement, "At Earth Fare, we have the highest quality standards for the health and safety of our food. If an ingredient is harmful to our bodies, it's banned from our shelves," I almost blew a sprocket. Because it's not. It's just added as celery juice powder instead of the pure chemical, which is more regulated and more accurately added. So, the next time you're buying your all-natural bacon, that reads "no nitrates or nitrites are added," make sure to look for the caveat, "other than those which naturally occur in celery powder."

ADDING SALT AND SUGAR

Salt, the other ingredient in the curing mixes, helps to dry the meat as well as add flavor and provide additional protection from pathogenic bacteria. Whereas lactic acid bacteria are present in meat, they are greatly outnumbered by pathogenic bacteria. Adding salt in the correct concentration (usually around 2–3.5% of the weight of the meat) suppresses the pathogenic bacteria while allowing the salt tolerant lactic acid bacteria to grow. Sugar may be added, too, typically dextrose (usually in the 0.5–2% range), which can potentially provide sweetness or at least balance the saltiness in the cure but is also added to provide a source of energy for the lactic acid bacteria doing the fermenting. The addition of sugar aids the fermentation process, which leads to greater organic acid production (and flavor), and it also aids in dropping the pH to provide more sourness and inhibit pathogenic bacteria.

REDUCING THE REDOX POTENTIAL

In fermenting meat, the redox potential, also known as the oxidation-reduction potential, refers to the measurement of the potential for electrons to be transferred between molecules involved in the fermentation process. Say what? OK, bear with me here. Redox, or oxidation-reduction, is just a means of tracking the flow of electrons in chemical reactions. Electrons, those negatively charged subatomic particles that surround the nuclei of atoms, can be shared among atoms, thereby forming bonds and making molecules. When electrons move between atoms in molecules, bonds can be broken and new bonds formed. If one atom or molecule gains an electron, it has been reduced, which also means that another atom or molecule must have lost that electron and has been oxidized. I teach my students the mnemonic LEO-GER—lose electrons oxidation, gain electrons reduction. You probably already take an antioxidant supplement or three (vitamin C, vitamin E, resveratrol, quercetin, the list goes on), but have you thought about what that means? Antioxidants are compounds that themselves can be easily oxidized (lose electrons), thereby protecting your own cells from oxidation. So, your body, like most ferments, enjoys an environment with a low redox potential, which is more protective against oxidation, a process that can lead to cell damage and death.

Redox potential plays a critical role in determining which microorganisms thrive and which ones do not during the fermentation process as well as which compounds are produced in the process. During meat fermentation, various microorganisms, including bacteria and fungi, use the available nutrients and substrates in the meat to carry out metabolic reactions that result in the production of organic acids, flavor compounds, and other products. The redox potential in fermenting meat can vary depending on several factors, such as the type of microorganisms present, the pH of the meat, the availability of oxygen, and the presence of antioxidants. Generally, a lower redox potential is more favorable for the growth of certain microorganisms involved in meat fermentation, such as lactic acid bacteria. By contrast, a higher redox potential may promote the growth of undesirable microorganisms, such as spoilage bacteria.

Measuring the redox potential in fermenting meat can help to optimize the fermentation conditions, control the growth of microorganisms, and ensure the quality and safety of the final product. However, this is something that is generally done only in commercial meat production. At the same time, certain factors can be controlled, even by the home fermenter, to help ensure a safer and tastier fermented meat product:

1. Oxygen exposure: The presence of oxygen can increase the redox potential in meat, so limiting its exposure is important. This can be done by using vacuum packaging or other airtight containers.

2. Temperature control: The temperature during fermentation can affect the redox potential of meat. Lowering the temperature can slow down the fermentation process and reduce the redox potential.

3. Salt concentration: Salt is used in meat fermentation to control the growth of bacteria and mold. However, a salt concentration that is too high can increase the redox potential, so it is imperative to use the correct salt concentrations for the desired product.

4. pH level: The pH level of the meat can also affect the redox potential. A more acidic environment can help reduce the redox potential. This can be achieved by adding acids like vinegar or citric acid to the meat or simply by monitoring the pH of your product to ensure it's in the correct range.

Overall, by controlling these factors during the fermentation process, it is possible to reduce the redox potential in fermenting meat.

INTRODUCING LACTIC ACID BACTERIA

You do not need to inoculate your meat with lactic acid bacteria because they are already present in the meat. However, I would highly recommend doing so with a known bacterial starter culture. This not only ensures that you have the proper bacteria to ferment your product but allows you to select for certain bacteria that will provide particular flavors and aromas to match the desired product. Different cultures will also ferment more effectively under different conditions, so it makes sense to choose a starter culture to suit your particular environmental needs. Additionally, many starter cultures now have bacteria that produce bacteriocins, compounds that inhibit the growth of other microbes, to protect the meat from pathogenic bacteria.

LOWERING THE PH

The initial fermentation step, which precedes the much longer curing process, drops the pH to a minimum value that will likely increase slightly during curing. The pH will drop from roughly 5.7 at the start of the ferment to a value dependent on the bacteria involved, the temperature of the ferment, and the amount of additional sugar added. The time of the ferment also varies depending on the bacteria and the temperature of the ferment. The higher the temperature, the faster the ferment will be completed. A high temperature ferment (at least 37°C, body temperature for you Fahrenheit adherents) can finish in as little as 12 hours and reach a pH as low as 4.2. More moderate fermentation temperatures, closer to 24°C (that's 75°F, which I will henceforth refer to as °American, since we are pretty much the only country in the world that insists on using this temperature scale), will result in a longer fermentation time (up to a few days) and more moderate pH values of 4.6–5.0. However, unless sugar is added to the meat, the pH value will generally not fall below about 5.0. Sugar is the key ingredient in achieving the lowest possible pH value for the range that is determined by the bacteria and fermentation temperature. Sugar will not impact the fermentation time, though. Fermentation time is temperature dependent. Following the fermentation phase, the pH will generally increase slightly to an equilibrium value during maturation. In fully dry, long maturing sausages, the pH can end as high as 5.6 or even 5.8. I start to freak out a little if the pH gets above 5.3, which really is uncommon unless you're aiming for a true southern European-style sausage; but another, arguably more important variable that must be considered from a food-safety perspective is the water activity (Toldra 2012).

DECREASING THE WATER ACTIVITY

The water activity (a_w) is defined as the ratio of the vapor pressure of the water in the meat product under stable conditions with the surrounding atmosphere to the vapor pressure of distilled water under identical conditions. Simply put, the a_w refers to the amount of free water in the meat that is available for microbial growth and chemical reactions. Let's say your product has an a_w of 0.90. That means that the vapor pressure of the water in the meat is 90% that of distilled water in the same conditions. Fermented meats have a reduced water activity due to the removal of water during the curing and drying process, which inhibits the growth of spoilage and pathogenic microorganisms. The question is, What is a low enough value? And that is the key question for determining whether your product is ready for consumption (when referring to fermented meat products that are not meant to be cooked, but rather consumed "raw"). For dry sausage that does not need to be cooked (salami, pepperoni, etc.), the a_w should be in the 0.85–0.86 range. For semidry products like summer sausage, an a_w of 0.92–0.94 is what you're shooting for. The problem with a_w is that it's not overly straightforward to measure. You can pick up a "cheap" a_w meter in the $500 range, which is likely out of the range of most home fermenters. These bargain-basement a_w meters are meh at best and may give you fairly accurate values … occasionally. To be able to measure a_w more accurately and on a consistent basis, you really need to look in the $5,000–8,000 range (definitely out of the range of home fermenters). Fortunately, there is a fairly easy workaround that doesn't require a direct measurement of a_w. By measuring the mass of your product prior to fermentation and maturation and then again when you think it may be ready, you can determine the amount of moisture that has been removed from the product. For whole-muscle ferments that are fully dried and ready to eat (country ham, capicola, etc.), the product is ready when 40–50% of the total weight has been lost. For dry sausages that are ready to eat, shoot for a 30–40% weight loss to determine when they're done.

SMOKING AND MOLD

There are two ways to protect the casing of your sausages or the outer layer of your whole-muscle ferments—smoke or mold. If the product is to be smoked, this is typically done either in conjunction with the fermentation phase (if the temperature requirements for both align) or following the fermentation. If mold is to be used to protect the casing, then the mold can typically be grown during the fermentation phase (again, as long as the temperature ranges of the bacterial culture and the mold culture overlap —the humidity will just need to be increased for the mold, but this doesn't impact the bacterial ferment). Smoke dries the meat, similar to salting, which helps to create an unsuitable environment for pathogenic bacteria. However, the curing process will draw far more moisture out of the meat over a long period of time. More importantly, the smoke itself coats the surface of the meat or casing with antimicrobial and antioxidant compounds that protect the meat from pathogenic bacteria and molds and also prevent oxidation of the fat. Not to mention the fact that smoked meats taste amazing! Smoke can be used for sausages in casings or for whole-muscle ferments, which is an added benefit. By contrast, intentionally inoculating your sausage casing with mold is another great way to protect your product and provide additional flavor compounds. Mold can also be used for whole-muscle ferments that have been stuffed into a casing, such as a bladder or stomach. I just wouldn't recommend using mold to protect your ham that is hanging unsheathed, if you will. By far, the most common mold to use is one of the many *Penicillium* species available for meat ferments. This is the beautiful, snowy white mold you've probably seen on salamis that are hanging and curing. By intentionally inoculating your product with mold, you protect it from any spoilage molds that might otherwise grow on the product. Additionally, depending on the mold, the enzymes excreted by the mold can help to break down carbohydrates, proteins, and fats in the meat, thereby leading to greater flavor development. We've been using a lot of *Aspergillus* mold recently in our lab as well. Whereas the *Aspergillus* doesn't tend to coat the casing nearly as well as *Penicillium* and tends to sporulate more easily, leading to some funky looking mold growth, the flavor that develops from the *Aspergillus* is superior in my opinion. It's often subtle, but the umami character is increased from the use of *Aspergillus*, the result of protein degrading enzymes in the mold that lead to more individual amino acids and umami taste. The fact that both methods of protecting the outer layer of your meat ferment—whether smoke or mold—also increases the flavor potential, is a huge win in my book. Table 7.1 provides the recommended concentration ranges of key ingredients used in curing and fermenting meats.

TABLE 7.1. Key ingredients and their amounts for fermented meats

Ingredient	Meat weight (%)
Salt	2–3.5 (the longer the overall maturation time, the higher the salt concentration)
Curing salt (#1 for sausage meant to be cooked, #2 for fully dry sausage)	0.25
Sugar (preferably dextrose)	0.5–2 (the more sugar, the lower the minimum pH)
Spices	0.1–2 (a loose range, depending on sausage type and personal preference)

Chapter 8

PRESENT

Just a couple of decades ago, you would have been hard pressed to find any type of fermented meat product in a grocery store in this country other than some mass-produced salami and pepperoni products. Nowadays you can walk into any grocery store in most parts of the country and head over to the specialty products area where, complementing the veritable sea of cheeses, you are likely to find a fermented meat section that rivals that of the cheeses. And many of them may even be locally produced. Go to a high-end butcher shop and you'll find even more options. The current fermented meat landscape is pretty healthy and looking better every day. Chefs all over the South are integrating fermented products into their dishes these days. Some of the noteworthy ones include Sean Brock, formerly of Husk in Charleston and now of Audrey and June in Nashville; Patrick O'Cain of Gan Shan West in Asheville (and formerly Gan Shan Station, also in Asheville, which sadly closed in 2020); and Misti Norris, chef/owner of Petra and the Beast in Dallas.

Chances are, if you haven't been living under a culinary rock for the last decade or so, that you've heard of Sean Brock. The James Beard award-winning chef and author has pioneered and, some would say, perfected the not-so-niche-anymore upscale southern cuisine genre. He sources local, heritage breed animals and heirloom plants for his traditional southern recipes with innovative, modern twists. His new restaurants, Audrey and June, incorporate all sorts of fermented delicacies, including meats like the 18-month country ham and summer sausage. They even employ a fermentation specialist, Elliot Silber, who has both a culinary and a food chemistry background. His official title is R&D and Fermentation Lab Manager, and his focus is on developing new products and ingredients as well as on using what would otherwise be waste products and turning them into delicious additions to the culinary experience.

Patrick O'Cain's restaurant Gan Shan Station opened in Asheville in 2015 and focused on Asian comfort foods using locally sourced ingredients from southern Appalachia. He opened the sister restaurant, Gan Shan West in 2017, which was also a hit. O'Cain is notable for his early adoption of koji curing meats. Gan Shan is one of the first restaurants in the region that I am aware of to use koji to "dry age" meat and to use shio koji—the quick ferment made from adding water and salt to koji and allowing it to ferment for about a week—to marinate meat. The enzymes in the koji break down the proteins in meat, not only tenderizing it but also producing plenty of umami character from the resulting amino acids. Coating meat in koji can also be used as a much faster substitute for dry aging, as the mycelium draw moisture out of the meat, cutting the process down from weeks to days, with very similar results.

Misti Norris is the chef/owner of Petra and the Beast in Dallas, which focuses on whole-animal butchery and the use of locally sourced ingredients. Fermentation plays a big role in the dishes served at Petra and the Beast, adding to the unique flavors as well as making use of materials that might otherwise be considered waste. Being a pioneer is not always easy, however. Fermented foods and flavors are relatively new to the Texas consumer, and it took a bit of time for customers to learn to trust and accept some of the fermented delicacies. The fact that they were delicious probably helped a bit, but it wasn't just the customers who had to be won over. Apparently, it took some time for health inspectors to educate themselves regarding the safety of the foods being prepared at Petra and the Beast. Before that time, however, there were instances when the inspectors poured bleach over the fermented foods and threw them in the trash because they were sitting out at room temperature, fermenting. It is probably somewhat alarming to a health inspector the first time they see a piece of meat growing mold (koji) at room temperature. I have no doubt that convincing them that it was perfectly safe took plenty of time and patience.

Whereas the three chefs covered have clearly influenced the current meatscape of the South, there are a couple of other influential players whom I should mention. I am not ashamed to admit that I am a total fanboy of farmer, educator, and author Meredith Leigh. She is pretty much

my meat fermentation idol. Leigh has degrees in Environmental Science and Sustainable Agriculture, has owned and managed farms as well as a restaurant and butcher shop, consults, hosts workshops and courses, and is an award-winning author of *The Ethical Meat Handbook* (2020) and *Pure Charcuterie* (2018). She takes a holistic view on farming, animal husbandry, and the preservation of foods through fermentation to create a sustainable food system. Not only has she preserved and reinvigorated many old-school techniques for fermenting and curing meats, she has also experimented extensively with koji curing of meats. She also helps lead Carbon Harvest, a regional carbon-offset program that works with landowners to reduce their carbon footprints and attract investment to sequester additional carbon in the region. And then there's the Fermentation School, which she also co-leads, that offers classes in fermentation and supports independent women educators. I have no idea how she has time to do everything that she does, but I'm grateful for everything that she does do. Rock on, Meredith!

Finally, I couldn't discuss the current meatscape and not cover one of the largest producers of country ham in the country, which also happens to be right in my backyard. Goodnight Brothers Country Ham in Boone, North Carolina, has been curing hams since 1948. As early as the 1930s, the Goodnights were in the produce distribution business, selling produce off the mountain that they had purchased in the High Country. Following World War II, all the brothers except for JC, moved down to Charlotte to expand their business. JC stayed in Boone, continuing to sell produce and also curing country hams. The business in Charlotte eventually merged with another company, but Goodnight Brothers lived on in Boone, where the focus became country ham. A new facility was opened in 1967, and in 1970, JC hired his son, Jim, to run the country ham business. In 1985, Jim brought in his first cousin, Bill Goodnight, to help with the business. They continued to expand it and moved into another new facility, their current location, in 1999. They now have three locations in Western North Carolina and distribute their products all over the country. Their country ham is salty, rich, and delicious. The next time you see it in your local store, make sure to pick some up. You will not be disappointed!

Chapter 9

FUTURE

To pick up where I left you hanging in chapter 4, let's discuss my crystal ball visions of the future of meat.

CULTIVATED MEATS

The first and most important prediction is cultivated meats. While these are not strictly produced using a fermentative process, the technologies are very similar, and cultivated meats are one of a few types of emerging alternative proteins, a broader field in which fermentation plays a large role. So, a little leeway is appreciated in order to describe what I see as arguably the most important advancement in meat production in the coming years. Cultivated meats are produced by growing the animal cells in bioreactors (think tricked out fermenters with lots of inputs and outputs). Stem cells are taken from the animals, placed into bioreactors, and force-fed a dense nutrient solution to encourage efficient growth, not unlike what happens inside the animals' bodies. Those stem cells, triggered by different growth factors, begin to differentiate into the various types of tissues normally produced in the animal, including muscle tissue. Once the differentiated tissues are grown, they are then harvested for various applications. The muscle tissue can be sold and thrown on the grill just like any other steak that would otherwise be harvested from an animal (although, admittedly, beef is harder to produce in this manner with its high fat content).

It may sound pretty sci-fi, in vitro steak, but it's reality. As of 2022, there were 150 companies around the world producing cultivated meat. Most have not hit the market yet, but there is a cultivated chicken product that is already available for sale in Singapore, and there are several more that should be getting there in the next year or two. In fact, in June 2023, the US Department of Agriculture (USDA) approved the production and sale of laboratory-grown cultivated meat in the United States, making us the second country after Singapore to do so. Arguably, these are the alternative protein sources that may have the greatest initial consumer appeal, since it's actual beef or chicken or fish, just grown in vitro rather than in vivo.

There are real concerns with this type of alternative protein regarding its environmental impact, however. If, for example, cultivated beef is grown using current pharmaceutical methods and pharmaceutical-grade inputs, the meat could actually have a larger climate impact than the cattle-grown beef. No doubt there will be efficiencies gained from scaling up production as well from advances in the production methods themselves. Also, if food-grade inputs were to be used instead, these would be less expensive and energy intensive than their pharmaceutical-grade counterparts, which could cut down on the climate impact. Time will tell, however. And there are the political hurdles as well. Florida became the first state to ban cultivated meats in May 2024, an act undertaken in large part to protect the livestock industry in the state, whose members are, understandably, generally opposed to cultivated meats.

Perhaps bioreactor burgers aren't your thing. I have yet to try cultivated meat, so I couldn't tell you if there is a discernible difference from the "real" thing.

DETOXIFYING FOODS

For my next future prognostication, I think I'll also continue on with a thought from chapter 4. That is, using fermentation as a means of detoxifying certain foods to make them edible. What could this possibly mean when it comes to meat, you ask. Well, I'm not talking about our version of *hákarl*, the fermented Greenland shark (which is a national dish of Iceland, although most Icelanders will tell you to stay very far away from it) if that's what you're afraid of. Rather, my focus here is on alpha-gal syndrome, the allergy to mammal meat that people develop after being bitten by a Lone Star tick, although other ticks likely cause the same allergy as well. It is called alpha-gal syndrome because it's an allergic reaction to the sugar molecule, galactose-α-1,3-galactose, which is found in the cell membranes of nonprimate mammals. When a tick feeds on a nonprimate mammal and then feeds on a primate mammal (like you), it can transfer some galactose-α-1,3-galactose molecules to your bloodstream. In some people, this causes an overactive immune response to the foreign molecule, and a cascade of antibodies is produced, hence the allergic reaction. The immune response remains as well, so that the next time the person eats mammal meat, they get the same allergic response, which in some people can cause anaphylaxis.

It's unclear why some people develop the allergy and others do not; nor is it clear why it's more severe in some than in others. It can go away after some time (again, for unknown reasons), but it can also get worse with subsequent tick bites. In short, it sucks, and I'm grateful that I haven't gotten it. But, I do wonder if some fermented meat products might be safe to eat for people with alpha-gal syndrome. There are some known lactic acid bacteria and fungi that can metabolize galactose-α-1,3-galactose, so would it be possible to introduce these microbes to the fermenting meat to metabolize the sugar? Are there other, more commonly used meat fermenting bacteria that may also metabolize the sugar sufficiently? I don't know yet, but I do intend to conduct research to find out.

ASPERGILLUS FUNGI IN MEAT PRODUCTS

Along similar lines (but only because *Aspergillus* is one of the fungi that can metabolize galactose-α-1,3-galactose), another future that I see for the meat fermentation industry is a much greater use of *Aspergillus* in the production of meat products. Don't get me wrong. There are plenty of innovative chefs and butchers out there already using the fungus for the production of some delicious meat products. Most of it so far, however, has been in the use of koji (*Aspergillus* on a substrate like rice, wheat, or barley) to simulate a dry-aging effect in rapid order. The meat is held at a high temperature and humidity for a couple of days to encourage the mold growth. As the mold grows and covers the meat, the mycelium works its way into the meat and not only dries the meat as it removes moisture to support the fungal growth but also excretes enzymes that break down the fat and protein. The enzymatic activity is what gives the meat a little additional umami boost from the amino acids that make up the proteins. However, the process typically lasts only a couple of days before the mold is removed and the meat is prepared.

However, a much greater potential could be tapped into were the koji to be used for a longer aging process. We have begun to coat some of the sausages and whole-muscle ferments in koji rather than using smoke or the more common mold for this task, *Penicillium*. The *Penicillium* coats the casings around the meat thoroughly, which is great, as it protects the casings from any unwanted mold or yeast growth. That said, it doesn't introduce too much from a flavor perspective to the aging meat. *Aspergillus*, by contrast, introduces a ton of additional umami character to the meat, which is great; but it doesn't coat the casings as thoroughly, and it tends to sporulate. This leaves the possibility open for unwanted mold growth during the long aging process. If we can dial in the proper environmental conditions for more thorough *Aspergillus* growth without sporulation, I think we'll have a truly viable new option for the meat industry.

Chapter 10

RECIPES

—

You can ferment meat with no starter culture, as the requisite microbes should already be present. However, it's not guaranteed, and there are pathogenic bacteria present as well, so it's less certain that the fermentation will go as planned. My advice is to use a starter culture to better ensure that things go as intended. There are plenty of sources for meat starter cultures. I am a fan of a few companies for my meat fermentation needs: Butcher & Packer (www.butcher-packer.com), The Sausage Maker (www.sausagemaker.com), and Craft Butcher's Pantry (https://butcherspantry.com/). All are great sources for starter cultures as well as for casings and equipment.

....

KOJI-CURED STEAK

Prep time: 15 minutes | Fermentation time: 36 hours
Yield: However many steaks you used

When it comes to using koji with meat, there are a couple of options. One involves using shio koji as a quick and effective marinade, as described in chapter 8. The application described here is more of a quick cure or accelerated dry aging with an extra kick of umami. It's really quite quick when evaluated against true curing or dry aging. The issues that may be difficult for some are the elevated temperature used with raw meat and the fact that you are growing mold on your meat. Feel free to use it with other meats as well, but experiment with small amounts first. It works great with red meat but can be more hit-or-miss with poultry and seafood. If it ends up smelling foul at the end of the process, something went wrong, and you should not eat it. Trust your nose on this one!

Ingredients

- Ribeye, NY Strip, or any other delicious, fatty cut of beef (thick cuts work best or they can potentially dry out too much)
- Salt (1.5% or enough to fully coat the steak)
- Sugar (1.5% or enough to fully coat the steak)
- Barley koji (3% or enough to fully coat the steak)—barley koji produced from red rice koji works great, but barley koji made from *A. luchuensis* could be fun for a citrus kick

Instructions

Weigh the steak and calculate 1.5% of that weight for the amount of salt and sugar to use. Calculate 3% of the weight for the amount of barley koji to use.

Mix the salt and sugar together and rub the steak with mix. Be sure to cover the entire steak. The 1.5% should get you close to the right amount, but you may use less or more to coat the meat.

Place the meat on a rack over a pan at room temperature to allow for airflow and the drainage of juices from the meat.

Leave the meat coated with the salt and sugar for about 1 hour. The salt draws moisture out of the meat, which will enable the koji to better adhere when it's added; and the salt also provides some protection from pathogenic bacteria. The sugar provides a little fuel for the koji to get going and start growing as quickly as possible.

While the meat is resting with the salt and sugar for 1 hour, grind your barley koji to a powder.

After 1 hour of resting with the salt and sugar, add the powdered koji to the steak. Make sure to completely cover it and rub it in well.

Place the meat in your fermentation chamber on a rack with a pan underneath for drainage and culture it at 30°C (86°F) and 90% humidity for about 36 hours. At this point the meat should be fully coated in mold and smell of mushrooms, soy sauce, and possibly a little ammonia. It should not smell like rot or decay, though. If it does, toss it, order some takeout, pour yourself a fermented beverage, and chalk it up to a failed experiment.

The meat can be cooked at this point or it can be stored in the refrigerator for a few days until ready to use. Before cooking, scrape the mold off and pat the meat dry. You really don't need to season it any more, but if you do, do not add any more salt. It will be plenty salty at this point.

Fermentation

As the *Aspergillus* grows, the mycelium covers the meat and even works its way into the interior of the meat to a degree. To fuel the growth, enzymes are excreted from the mycelium to digest material in the environment, thereby producing nutrient molecules that are taken up and used. In this case, the enzymes are mainly degrading the proteins and fats from the meat, producing umami rich amino acids (primarily glutamate/glutamic acid) and fatty acids, respectively. Besides the umami character produced from the amino acids, the fatty acids produced from the degradation of the fats are responsible for quite a bit of flavor as well, specifically sour, funky flavors. Additionally, the outer layers of the meat are dried out as moisture is drawn from the meat by the action of the salt and the mycelium. As noted earlier, the flavors and texture produced in 1 day and a half are not dissimilar from those generated in a long dry-aged piece of meat.

....

SMOKED ANDOUILLE

Prep time: 90 minutes | Smoking time: 6–8 hours | Yield: 7 kg

OK, so technically this is a fresh sausage, not fermented, but it does have a couple of fermented products in it, and it's a good sausage for learning the overall process. Plus, it's delicious, and you can snack on it while you wait for your fermented sausages. In our lab, we used some garlic that we had fermented in honey (a very straightforward lactic acid ferment that involves 2 ingredients, honey and garlic, and time), but feel free to use unfermented raw garlic instead. Likewise, we used our Carolina Common beer, but you can use any beer you choose. If you want a substitute that is similar to the Carolina Common, look for an amber ale that isn't too hoppy, a Märzen, or an Extra Special Bitter (ESB).

Ingredients

- 7 kg pork shoulder
- 70 g fermented honey garlic (or raw garlic)
- 28 g cayenne
- 42 g black pepper
- 14 g thyme
- 1.5% (of the weight of the meat) salt (105 g)
- 700 g Carolina Common beer
- 0.25% (of the weight of the meat) cure #1 (17.5 g)

Instructions

Cut the meat and fat into 2.5 cm cubes and freeze.

Chill your grinder equipment in the freezer for a couple of hours prior to grinding to make the grinding easier and to keep the meat cold during the process.

Thaw the meat about halfway before grinding. The meat should have fairly soft edges, but still have a hard, frozen center.

Mix the meat and all other ingredients, including the curing salt, and grind with a ⅜″ plate.

119

Let the ground mixture sit in the refrigerator, covered, for 24–48 hours (this allows the sausage to bind and hold together better).

Stuff in beef casings (38–42 mm), although hog casings of a similar size would work well too (in fact, I prefer hog casings, as they're not as tough as the beef ones).

Poke lots of holes in the sausage using a proper sausage pricker or a needle (this prevents air pockets from forming between the sausage and casing as the sausage dries).

Hang and dry the sausages at room temperature for 1 hour.

Smoke at 60°C (140°F) for 2–3 hours with the wood of your choosing. We used pecan wood, which was delicious, but apple or cherry would work well too.

Raise the temperature 6°C (10°F)/hour to 77°C (170°F) and smoke the sausages until the internal temperature reaches 68°C (154°F).

Remove sausages from the smoker and spray them with water to cool the internal temperature to 49°C (120°F); dry them, package, and refrigerate until you're ready to enjoy.

These are cooked and ready to eat, but you'll probably want to reheat them when you're incorporating them into a dish.

....

CHORIZO

Prep time: 90 minutes | Fermentation time: 2 days
Smoking time: 6 hours | Yield: 15 kg

Ingredients

- 11.4 kg pork butt
- 3.6 kg pork belly
- 181 g garlic cloves
- 100 g pasilla pepper powder
- 13 g chipotle powder
- 37 g chile Colorado powder

- 30 g oregano
- 2% (of the weight of the meat) dextrose (300 g)
- 3% (of the weight of the meat) salt (450 g)
- 0.25% (of the weight of the meat) cure #2 (37.5 g)
- 0.025% (of the weight of the meat) SafePro® B-LC-007 (3.75 g)

Instructions

Cut the meat and fat into 2.5 cm cubes and freeze.

Chill your grinder equipment in the freezer for a couple of hours prior to grinding to make the grinding easier and to keep the meat cold during the process.

Thaw the meat about halfway before grinding. The meat should have fairly soft edges, but still have a hard, frozen center.

Mix the meat and spices and grind with a ⅜" plate.

Hydrate the bacterial culture according to the instructions on the package.

Add the hydrated bacterial culture and curing salt to the ground meat and fat and mix well.

Let the mixture sit, covered, in the refrigerator for 24–48 hours.

Stuff the mixture into 38–42 mm hog casings.

Poke lots of holes in the sausage using a proper sausage pricker or a needle (this prevents air pockets from forming between the sausage and casing as the sausage dries).

Hang the sausage in a fermentation cabinet at 22°C (72°F) and 85% humidity to ferment for 2 days.

After 2 days, cold smoke the sausage at 24°C (75°F) with pecan wood for about 6 hours.

Put the sausages back into the fermentation cabinet at 16–18°C (60–65°F) and 65–75% humidity.

Gradually reduce the temperature (1–2° every other day or so) to 13°C (55°F) and gradually bring the humidity down (1–2% every other day or so) to 65% humidity, if not there already.

Pro tip: weigh some (or all, if it's not too daunting) of the sausage after stuffing. If you do some, make sure to remember which ones they are. You can periodically weigh the selected sausages to determine how much moisture (and weight) they have lost.

Let the sausages cure and dry until they have lost 30–40% of their original weight.

Fermentation

You'll notice that I use a couple of different starter cultures in these recipes, and for each I'll give a brief overview of the culture, the fermentation, and the finished product. I've got to say that the B-LC-007 is one of my favorites. This is a fast fermenter, as in a 1- to 2-day fermentation time in a very accommodating temperature range (20–24°C / 68–75°F). It's a mixed culture of bacteria and yeast, with the lactic acid bacteria *Pediococcos acidilactici*, *Staphyloccocus carnosus*, *Staphylococcus xylosus*, *Lactobacillus sakei*, and *Pediococcus pentosaceus*, and the yeast *Debaryomyces hansenii*. The *Pediococcos acidilactici* produces lactic acid and is a protective species, as it produces pediocin, a bacteriocin with antimicrobial activity against *Listeria*, one of the main pathogens of concern in meat fermentations. The *Lactobacillus sakei* and *Pediococcus pentosaceus* are primarily responsible for a fast drop in pH through the production of lactic acid, while not producing an overly sour finished product, resulting in a taste that is much more like a traditional European sausage. The *Lactobacillus* also produces bacteriocins, providing additional protection against *Listeria*, whereas the *Pediococcus* produces enzymes to help degrade the proteins in the meat, leaving behind flavor-positive amino acids. The *Staphyloccocus* bacteria in the culture do a lot of heavy lifting. They produce enzymes that degrade both proteins and fats, producing a lot of the characteristic flavor of the sausage. Additionally, they reduce nitrate from the cure to nitrite, providing microbial protection as well as color development through the complexing of the nitrite with the myoglobin in the meat. On top of that, they reduce residual nitrite in the meat, thereby leaving behind no more than necessary. Finally, the *Debaryomyces* reduces overall acidity and inhibits rancidity through its lipid-degrading enzymes, while also adding to the finished flavor profile from the fatty acids produced in the process. In general, I thought this culture was spot on for our version of a fermented chorizo.

WAKALIM

Prep time: 90 minutes | Fermentation time: 1–2 days | Yield: 5 kg

I love this sausage. This is a traditional, spicy Ethiopian sausage that is not unlike a spicy pepperoni, just with North African spices. It is typically made with all lean beef. We cheated a bit, though, as you'll see, since we added in a little bit of pork fat for that delicious unctuousness you expect from a sausage.

Ingredients

- 4.75 kg lean beef
- 0.25 kg pork fat
- 650 g sweet onion
- 25 g garlic
- 25 g ginger
- 125 g berbere spice
- 25 g black pepper
- 25 g cumin
- 12.5 g mitmita chili powder
- 200 g (about 200 mL) lemon juice
- 2% (of the weight of the meat and fat) dextrose (100 g)
- 3% (of the weight of the meat and fat) salt (150 g)
- 0.25% (of the weight of the meat and fat) cure #2 (12.5 g)
- 0.025% (of the weight of the meat and fat) SafePro® B-LC-007 (1.25 g)
- Bactoferm® Mold-600

Instructions

Cut the meat and fat into 2.5 cm cubes and freeze.

Chill your grinder equipment in the freezer for a couple of hours prior to grinding to make the grinding easier and to keep the meat cold during the process.

Thaw the meat and fat about halfway before grinding. The meat should have fairly soft edges, but still have a hard, frozen center.

Mix the meat and spices, including the lemon and curing salt, and grind with a ¼" plate.

Grind the fat with a ⅜" plate and mix with meat and spices.

Hydrate the bacterial culture according to the instructions on the package.

Add the hydrated bacterial culture to the ground meat and fat and mix well.

Let the mixture sit, covered, in the refrigerator for 24–48 hours.

Prepare your mold solution by mixing it into filtered water at a rate of 3 g per 200 mL of lukewarm water, allowing the solution to equilibrate for 12 hours, and then diluting it to 1 L with filtered water.

Stuff the meat mixture into 38–42 mm hog casings and dunk them in a 10% vinegar solution (we've been employing this technique with success, as it washes the casing of any stray bits of meat and adds a little microbial protection until the mold takes hold).

After dunking the sausages in the vinegar solution, dunk them in the mold solution.

Poke the sausages with lots of holes using a sausage pricker or needle and hang them in a fermentation cabinet at 24°C (75°F) and 75–90% humidity.

After 1 day, turn the temperature up to 26°C (80°F) to accommodate mold growth. You can also give the sausage a spray with the mold solution at this point.

Once mold is evident, gradually reduce the temperature (1–2° every other day or so) to 13°C (55°F) and gradually bring the humidity down (1–2% every other day or so) to 65% humidity.

Let the sausages continue to dry and cure until they have lost 30–40% of their original weight.

Fermentation

The Bactoferm® Mold-600 is a pure culture of *Penicillium nalgiovense*, the most common mold used in sausage production. It produces the characteristic white/gray full coating of mold that you often see growing on salami or other fully fermented sausage. It really does produce a beautiful, snowy white coating, as you can see in the photo below. Besides making the sausages beautiful (OK, so maybe I'm a little weird about my version of beauty when it comes to fermented foods), the mold also provides protection against unwanted molds and yeasts, helps with the uniform drying of the sausage, and adds some mushroom flavor to the finished product as well, not unlike the *Penicillium* on a bloomy rind cheese.

Beautiful (yeah, yeah, I know), snowy, white mold covering sausages that are hanging in front of a smoked Tom Thumb sausage.

SALAMI CALABRESE

Prep time: 90 minutes | Fermentation time: 1–2 days | Yield: 5.5 kg

Ingredients

- 3.78 kg pork belly
- 1.62 kg back fat
- 11 g black pepper
- 11 g paprika
- 22 g chili flakes
- 11 g fennel seeds
- 324 g red wine
- 11 g dextrose
- 11 g table sugar (sucrose)
- 3% (of the weight of the meat and fat) salt (162 g)
- 0.25% (of the weight of the meat and fat) cure #2 (13.5 g)
- 0.025% (of the weight of the meat and fat) SafePro® B-LC-007 (1.35 g)
- Bactoferm® Mold-600

Instructions

Cut the meat and fat into 2.5 cm cubes and freeze.

Chill your grinder equipment in the freezer for a couple of hours prior to grinding to make the grinding easier and to keep the meat cold during the process.

Thaw the meat about halfway before grinding. The meat should have fairly soft edges, but still have a hard, frozen center.

Mix the meat and spices, including the curing salt, and grind with a ¼" plate.

Grind the fat with a ⅜" plate and mix with the meat and spices.

Hydrate the bacterial culture according to the instructions on the package.

Add the wine and hydrated bacterial culture to the ground sausage and mix well.

Let the mixture sit, covered, in the refrigerator for 24–48 hours.

Prepare your mold solution by mixing it into filtered water at a rate of 3 g per 200 mL of lukewarm water, allowing the solution to equilibrate for 12 hours, and then diluting it to 1 L with filtered water.

Stuff the mixture into 38–42 mm hog casings, and dunk them in a 10% vinegar solution, followed by a dunk in the mold solution.

Poke the sausages with lots of holes using a sausage pricker or needle and hang them in a fermentation cabinet at 24°C (75°F) and 75–90% humidity.

After 1 day, turn the temperature up to 26°C (80°F) to accommodate mold growth. You can also give the sausage a spray with the mold solution at this point.

Once mold is evident, gradually reduce the temperature (1–2° every other day or so) to 13°C (55°F) and gradually bring the humidity down (1–2% every other day or so) to 65% humidity.

Let the sausages continue to dry and cure until they have lost 30–40% of their original weight.

MERGUEZ

Prep time: 90 minutes | **Fermentation time:** 1–2 days | **Yield:** 9.5 kg

Ingredients

- 9.22 kg lamb shoulder
- 100 g fermented honey garlic (or raw garlic)
- 50 g sweet paprika
- 50 g smoked paprika
- 50 g cayenne
- 50 g coriander
- 50 g black pepper
- 50 g cumin
- 100 g dextrose
- 3% (of the weight of the meat) salt (277 g)
- 0.25% (of the weight of the meat) cure #2 (23 g)
- 0.025% Bactoferm® F-RM-52 (2.30 g)
- Bactoferm® Mold-600

Instructions

Cut the meat into 2.5 cm cubes and freeze.

Chill your grinder equipment in the freezer for a couple of hours prior to grinding to make the grinding easier and to keep the meat cold during the process.

Thaw the meat about halfway before grinding. The meat should have fairly soft edges, but still have a hard, frozen center.

Mix the meat and spices, including the curing salt, and grind with a ³⁄₁₆″ plate.

Hydrate the bacterial culture according to the instructions on the package.

Add the hydrated bacterial culture to the ground sausage and mix well.

Let the mixture sit, covered, in the refrigerator for 24–48 hours.

Prepare your mold solution by mixing it into filtered water at a rate of 3 g per 200 mL of lukewarm water, allowing the solution to equilibrate for 12 hours, and then diluting it to 1 L with filtered water.

Stuff the mixture into 38–42 mm beef casings (you could also use sheep casings, but those are a smaller diameter, which would result in a faster curing time) and dunk them in a 10% vinegar solution, followed by a dunk in the mold solution.

Poke the sausages with lots of holes using a sausage pricker or needle and hang them in a fermentation cabinet at 26°C (80°F) and 75–90% humidity.

Once mold is evident, gradually reduce the temperature (1–2° every other day or so) to 13°C (55°F), and gradually bring the humidity down (1–2% every other day or so) to 65% humidity.

Let the sausages continue to dry and cure until they have lost 30–40% of their original weight.

Fermentation

This culture is another fast fermenter that is a mix of *Lactobacillus sakei* and *Staphyloccocus carnosus*. The fermentation temperature range 22–32°C (72–90°F) is a little higher than for the B-LC-007, but still very reasonable. It produces more acid than the B-LC-007 as well, so the resulting sausage is a little more sour. As such, it is more conducive to Northern European–style sausages but can still be used for Mediterranean style sausages. It's pretty much the OG of meat starter cultures and certainly a good go-to culture that makes very tasty sausage. That said, it does not provide the microbial protection that other cultures do through the production of bacteriocins. Another thing to note is you must use pure dextrose (glucose) with this culture and not table sugar (sucrose, which is a disaccharide of glucose and fructose), as this culture can ferment dextrose but not sucrose.

LANDJÄGER

Prep time: 90 minutes | **Fermentation time:** 2 days | **Yield:** 17 kg

This is a traditional German dried sausage that is enjoyed by hikers and hunters when they need to carry food that doesn't require refrigeration. It's typically made with beef and pork, but we figured since it's a hunter's sausage we should use venison to replace the beef (also because we had a bunch of venison in the freezer). We stuck with the pork to provide the fat in the sausage and added lots of spice to complement the venison and provide a pick-me-up when out on the trail. We also went with narrow diameter collagen casings to make them more like snack sticks.

Ingredients

- 9.5 kg venison
- 6.5 kg pork belly
- 160 g fresh garlic
- 160 g caraway seed
- 160 g ground coriander seed
- 320 g ground black pepper
- 320 g ground allspice
- 80 g celery seed
- 160 g dextrose
- 750 g (about 750 mL) sherry
- 3% (of the weight of the meat) salt (480 g)
- 0.25% (of the weight of the meat) cure #2 (40 g)
- 0.025% (of the weight of the meat) Bactoferm® F-RM-52 (4 g)

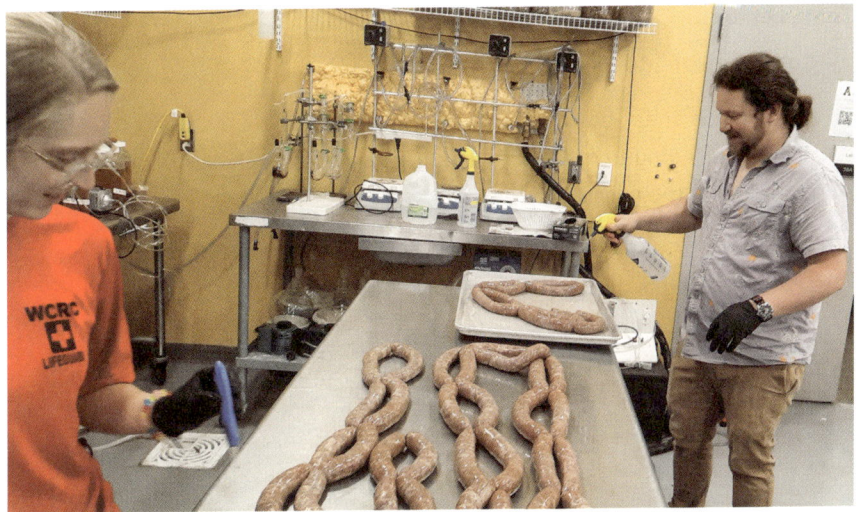

LeighAnne Baldwin poking the freshly stuffed sausages, and Daniel Parker spraying them with a *Penicillium* solution.

Some freshly stuffed sausages ready for the fermentation cabinet.

Don't eat them yet, Daniel!

Instructions

Cut the meat into 2.5 cm cubes and freeze.

Chill your grinder equipment in the freezer for a couple of hours prior to grinding to make the grinding easier and to keep the meat cold during the process.

Thaw the meat about halfway before grinding. The meat should have fairly soft edges, but still have a hard, frozen center.

Mix the meat and spices, including the curing salt, and grind with a ⅜" plate.

Hydrate the bacterial culture according to the instructions on the package.

Add the hydrated bacterial culture to the ground sausage and mix well.

Let the mixture sit, covered, in the refrigerator for 24–48 hours.

Stuff the mixture in 21 mm collagen casings.

Poke the sausages with lots of holes using a sausage pricker or needle and hang them in a fermentation cabinet at 26°C (80°F) and 85% humidity for 2 days.

After 2 days, cold smoke the sausages at 26°C (80°F) with apple wood for about 6 hours.

Put the sausages back in the fermentation cabinet at 16–18°C (60–65°F) and 65–75% humidity.

Gradually reduce the temperature (1–2° every other day or so) to 13°C (55°F) and gradually bring the humidity down (1–2% every other day or so) to 65% humidity, if not there already.

Let the sausages cure and dry until they have lost 30–40% of their original weight.

ASIAN SAUSAGE

Prep time: 90 minutes | Fermentation time: 1–2 days | Yield: 14 kg

Ingredients

- 9.5 kg pork loin
- 4 kg pork belly
- 135 g five spice blend
- 135 g candied ginger
- 135 g garlic
- 67.5 g chili crisp (try to use mostly crisp and not much oil)
- 135 g dark miso
- 1 kg (about 1 L) plum wine infused with rose petals
- 135 g dextrose
- 3% (of the weight of the meat) salt (405 g)
- 0.25% (of the weight of the meat) cure #2 (33.75 g)
- 0.025% (of the weight of the meat) Bactoferm® F-RM-52 (3.375 g)
- Bactoferm® Mold-600

Instructions

Cut the meat and fat into 2.5 cm cubes and freeze.

Chill your grinder equipment in the freezer for a couple of hours prior to grinding to make the grinding easier and to keep the meat cold during the process.

Thaw the meat about halfway before grinding. The meat should have fairly soft edges, but still have a hard, frozen center.

Mix the meat and spices, including the curing salt, and grind with a ⅜″ plate.

Hydrate the bacterial culture according to the instructions on the package.

Add the miso, wine, and hydrated bacterial culture to the ground sausage and mix well.

Let the mixture sit, covered, in the refrigerator for 24–48 hours.

Prepare your mold solution by mixing it into filtered water at a rate of 3 g per 200 mL of lukewarm water, allowing the solution to equilibrate for 12 hours, and then diluting it to 1 L with filtered water.

Stuff the mixture in 38–42 mm hog casings and dunk them in a 10% vinegar solution, followed by a dunk in the mold solution.

Poke the sausages with lots of holes using a sausage pricker or needle and hang them in a fermentation cabinet at 26°C (80°F) and 75–90% humidity.

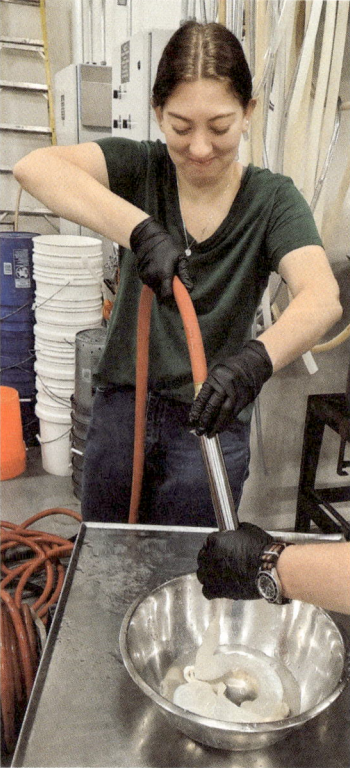

Top: Logan "Sausage Master" Isley about to go to work.

Bottom: The Sausage Master in her element.

But first she has to clean out the casings.

Once mold is evident, gradually reduce the temperature (1–2° every other day or so) to 13°C (55°F) and gradually bring the humidity down (1–2% every other day or so) to 65% humidity.

Let the sausages continue to dry and cure until they have lost 30–40% of their original weight.

....

TOM THUMB SAUSAGE

Prep time: 90 minutes | Fermentation time: 2 days | Yield: 8 kg

Ingredients

- 7.4 kg pork shoulder
- 30 g black garlic
- 81 g garlic
- 111 g ginger
- 37 g chili flakes
- 37 g sage
- 3.7 g nutmeg
- 15 g thyme
- 703 g Chocolate Salty Balls (the recipe for which is in chapter 20) or a beer of your choosing
- 111 g dextrose
- 3% (of the weight of the meat) salt (222 g)
- 0.25% (of the weight of the meat) cure #2 (18.5 g)
- 0.025% (of the weight of the meat) Bactoferm® F-RM-52 (1.85 g)
- Bactoferm® Mold-600

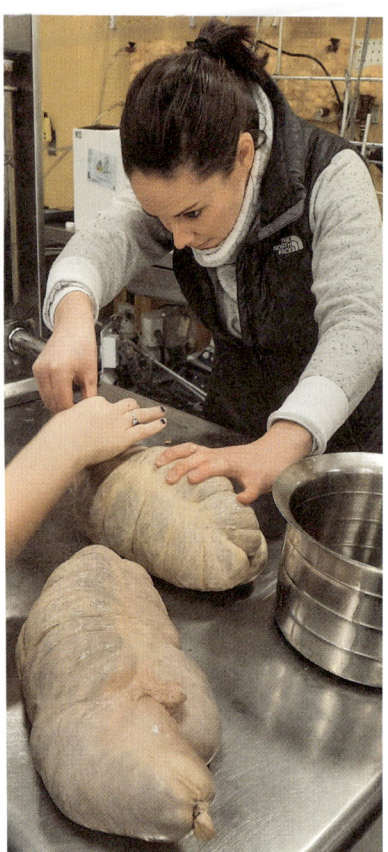

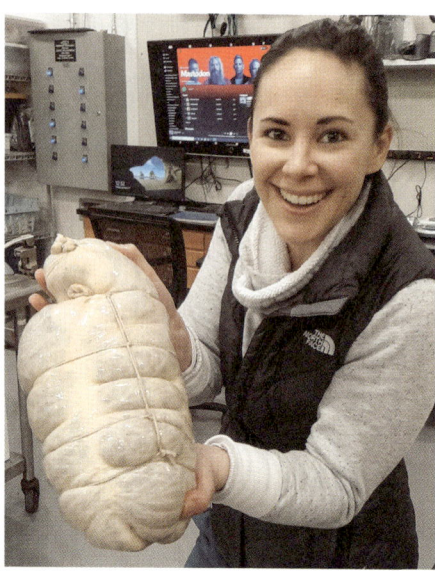

Left: Ali Turley filling and securing the Tom Thumbs prior to their fermentation and curing journeys. Aren't they just lovely?

——

Top: A proud mama holding her alien larva. Just kidding, a very happy Ali with a freshly stuffed Tom Thumb about to begin its fermentation journey.

All three versions (*from left to right:* coated in *Aspergillus*, smoked, and coated in *Penicillium*) of the Tom Thumb sausages after their fermentation journeys.

Instructions

Cut the meat and fat into 2.5 cm cubes and freeze.

Chill your grinder equipment in the freezer for a couple of hours prior to grinding to make the grinding easier and to keep the meat cold during the process.

Thaw the meat about halfway before grinding. The meat should have fairly soft edges, but still have a hard, frozen center.

Mix the meat and spices, including the curing salt, and grind with a 3/16" plate.

Hydrate the bacterial culture according to the instructions on the package.

Add the beer and hydrated bacterial culture to the ground sausage and mix well.

Let the mixture sit, covered, in the refrigerator for 24–48 hours.

Stuff the mixture into hog middle caps until the casing is pretty full and taut. It can handle quite a bit more meat than your average casing, so don't be afraid to overstuff that alien larval casing.

Dunk the sausage in a 10% vinegar solution.

Poke plenty of holes in the casing using a sausage pricker or needle after it has been stuffed and dunked.

Now you have options: (1) Dunk in the *Penicillium* solution (Bactoferm Mold-600); (2) coat in ground koji barley; or (3) smoke with cherry wood (or other wood of your choosing) at 26°C (80°F) for a minimum of 3 hours and up to a full day.

Hang the inoculated or smoked sausages in a fermentation cabinet at 26°C and 75–90% humidity.

Once mold is evident (if inoculated with one of the molds) or after 2 days (for the smoked version), gradually reduce the temperature (1–2° every other day or so) to 13°C (55°F) and gradually bring the relative humidity down (1–2% every other day or so) to 65% humidity.

Let the sausage ferment and dry until it has lost 30–40% of its original weight.

Part III

DAIRY

Chapter 11

PAST

—

The origins of fermented dairy predate recorded history, which means humans have been doing it for at least 7,000 years. It likely began sometime around the domestication of certain ruminants, like sheep, which happened around 8,000–10,000 years ago. That's a long time for us humans, which is not surprising. Milk is incredibly calorie and nutritionally dense, but it is also extremely perishable. It therefore stands to reason that once a means of preserving it was discovered, it would be widely used and passed on to every subsequent generation. It is also easily fermentable, and the process likely began with the happy accident of making the very first cheese. In fact, it's so easy that there are a few plausible scenarios that could have all led to the discovery.

The first, and most interesting in my opinion, involves early accessorizing by our ancient ancestors. We all have our preferences, whether it's a purse, a fanny pack, or a backpack. Why would that be any different several thousand years ago? The ancients needed convenient ways to carry their personal belongings as much as we do, maybe even more so. So, what were the options pre-Louis Vuitton? Well, the abomasa, the ruminant's fourth stomach, would certainly do in a pinch, especially when it's dried and inflated to hold the day's groceries, like the sheep milk you just collected to then carry to your ailing, 40-year-old grandmother in her ancient, decrepit state. By the time you made it to grandma's cozy cave, the milk had curdled, but in a good way, from the rennet found in the abomasa. Fortunately, the curds were still soft enough that grandma could gum them down, and if enough whey (and, importantly, lactose) was expelled from the curds, the result was more digestible as well. In fact, the rennet produced in the baby ruminant's gut makes the curds more digestible to them as well. At this point, it wouldn't have been hard for some curious ancient human to test the curdling theory in different carrying pouches until it was determined that the abomasa was the only one that did the trick.

The other scenarios, while equally plausible, are not nearly as exciting, in my opinion. Milk would, of course, curdle on its own, which, without refrigeration, happens pretty rapidly. If the curdled milk were salted, this would select for the nonpathogenic bacteria already in the milk and encourage a more favorable ripening process than the alternative, which could have been downright deadly. Before they knew it, people had a palatable and nondeadly early cheese product that would keep for a bit longer than raw milk.

A final scenario involves the addition of fruit juice to raw milk. Why not, right? The acid in the juice would curdle the milk, thereby producing more stable curds. Regardless of the method that was stumbled upon back in the day, all of the scenarios described above produced curds that were more stable than the raw milk itself, particularly if enough whey with lactose was expelled and if the curds were adequately salted. In fact, all three of these scenarios were probably stumbled upon at different times and in different places in history, which led to different cheese-making traditions in different locations throughout history. Which just goes to show how easy it would have been to stumble upon fermenting milk in the first place.

Now, fast forward several thousand years to the early American South, and that grand cheesemaking and dairy fermentation history slows to a near halt. It's not that early southerners had anything against fermented dairy, it's just that there weren't many cows, sheep, or goats in the South back in the day. Remember, pigs ruled the day back then. And still do to a large degree. Ever had pig's milk cheese? How about pig yogurt? Me neither. And there's a reason for that. Pigs are notoriously difficult to milk and don't produce a lot of milk, nor do they produce milk continuously. In one instance of extreme cheese making, Dutch farmer Erik Stenink, with ten of his closest friends, spent hours milking his sows to produce a few pounds of cheese. The cheese itself was purportedly pretty chalky and grainy, but the farmers' hard work did pay off. The cheese, called Piggie Palace, sold for $2,300 per kilogram, which, supposedly, is the most expensive cheese ever produced.

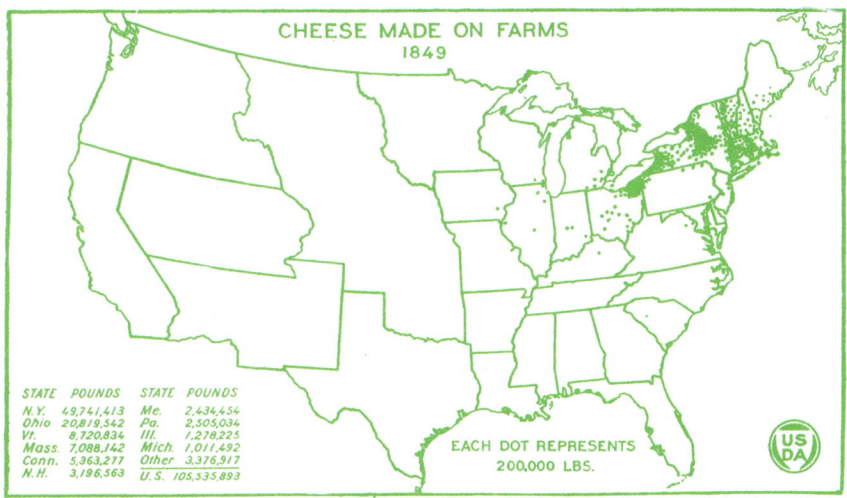

CHEESE MADE ON FARMS
1849

STATE   POUNDS   STATE   POUNDS
N.Y.   49,741,413   Me.   2,434,454
Ohio   20,819,542   Pa.   2,505,034
Vt.   8,720,834   III.   1,278,225
Mass.   7,088,142   Mich. 1,011,492
Conn.   5,363,217   Other 3,376,917
N.H.   3,196,563   U.S. 105,535,893

EACH DOT REPRESENTS
200,000 LBS.

MAP 11.1. USDA map of the contiguous thirty states showing cheese
production on farms in 1849. Downloaded from "Maps ETC" (USDA 1923),
courtesy of the Library of Congress.

So, the whole history of dairy fermentation in the early American South can be summed up as such—there really isn't one to speak of. Check out the USDA map of cheese made on farms in 1849 America (see map 11.1). Is that a dot in South Carolina? I'm pretty sure I see a dot each in Tennessee, Kentucky, and Virginia (which at the time also comprised West Virginia), but clearly cheese making in the South was not anything to write home about.

But what about your grandmas who churned butter for 12 hours a day and referred to buttermilk as milk and regular milk as sweet milk? I'm not saying there were no cows (or sheep or goats) in the South back in the day, and with the ones that were kept on small, family farms, their milk was certainly fermented for preservation purposes. But there really weren't enough of them to make any sort of rich tradition in dairy fermentation that is uniquely southern. Really, the only fermented dairy products that have any history in the South are hoop cheese and farmer cheese.

Hoop cheese is one of the simplest cheeses to make. It is made from cow's milk that is coagulated and has its whey expelled. The name comes from the round mold, open on both ends, in which the curds are placed to drain the whey. The curds are further pressed to completely expel the whey, and then the wheel of fresh cheese is wrapped in cheesecloth, parchment, or red wax, as is popular nowadays if you can find it. It is an unaged cheese

with reasonably high moisture content and little to no salt, so it is quite perishable. This, along with the fact that it has a pretty mild flavor, is the reason
why it is hard to find these days. Ashe County Cheese in West Jefferson,
North Carolina, makes a darn good version if you're interested in trying it.

Farmer cheese is a similar product, although it often has cream and salt
added to the milk. It is also unaged with a high moisture content, so also
quite perishable. I don't believe either type of cheese is officially recognized
by the American Dairy Association, and I've seen both referred to as pot
cheeses before, but that seems a stretch. Pot cheese, which is more similar to ricotta or queso blanco, is softer than either hoop or farmer cheese,
hence the fact that it was traditionally served in a pot. Quark in Austria
is sometimes referred to as *topfen*, which translates from German as pot.
And that pretty much sums up the history of dairy fermentations in the
South. Fortunately for us, the present is a lot richer in fermented dairy in
our beloved South.

BACKGROUND

—

Simply put, dairy fermentation makes use of microorganisms to transform milk into various products, such as yogurt, cheese, and kefir. These products have been a part of human diets for thousands of years and are still enjoyed by many people today. Notably, they all share one common characteristic. The fermented dairy products are far more stable than raw milk, thus allowing for the preservation of a nutrient-dense food item that would otherwise spoil fairly rapidly without refrigeration. Obviously, cheese is the most stable of the fermented dairy products, but all fermented dairy is more stable than the raw dairy.

The process of dairy fermentation begins with the selection of appropriate microorganisms, typically bacteria and/or yeast, that are added to the milk. These microorganisms then digest the lactose, the primary milk sugar, and produce lactic acid, which changes the pH of the milk and causes it to coagulate. In the case of yogurt, specific strains of bacteria, such as *Lactobacillus bulgaricus* and *Streptococcus thermophilus*, are added to the milk. The bacteria ferment the lactose in the milk, producing lactic acid, which gives yogurt its characteristically tangy taste and causes the milk to coagulate. Kefir, by contrast, is produced through the fermentation of milk using a combination of bacteria and yeast. The microorganisms in kefir produce a slightly thick and mildly tangy beverage that's more akin to a yogurt smoothie than to yogurt itself. Besides cheese, these are two of the more common dairy ferments that we are familiar with in the United States; but there are dozens of fermented dairy products from all over the world that vary depending on the source of the milk, the microbes involved in the fermentation, and the processes used. Some of note include *chal,* a camel's milk yogurt (nice and low fat, if that's your thing) from Turkmenistan; *filmjolk,* a notoriously slimy and ropy cow's milk ferment from Sweden; *kumiss,* a product from Russia that is similar to kefir, except it comes from horse's milk; *smetana,* which is a fermented heavy cream product like sour cream or crème fraîche from Central and Eastern Europe; and *quark,* which is a German product similar to cream cheese. Let's briefly discuss cheese production, though, as it is the most varied and complex and will also allow us to discuss some milk and cheese chemistry.

CHEESE PRODUCTION

Coagulation

The first step in the cheesemaking process is coagulation. This is the step whereby the primary milk protein, casein, and milk fat are separated as curds from most of the lactose and the secondary milk protein, whey, which are dissolved in the water. The key player in coagulation is the casein, and there are a few different processes that can be used to coagulate it in slightly different manners. Casein proteins assemble into large, spherical, macromolecular complexes referred to as casein micelles, which form a colloidal suspension in milk. These micelles are aggregates of thousands of individual casein molecules held together mainly by ionic bonds, the electrostatic attraction between two oppositely charged residues on the proteins. The main characteristic of these casein micelles that allows for coagulation is the "hairy" outer surface. It's not really hairy, but the outer protein surface, the κ-caseins, have carbohydrate side chains with negatively charged acidic groups that hang off the outer layer like nanoscopic hairs on the protein ball. This negatively charged, "hairy" surface layer is the key to unlocking coagulation.

Acid coagulation is the most straightforward means of coagulating the milk. When lactic acid bacteria are allowed to proliferate in warm milk (20–32°C, which translates to 68–90°American) and produce an excess of lactic acid from the lactose in milk, the pH of the milk drops. The pH is a measure of the positively charged hydrogen ions in solution. The more hydrogen ions, the lower the pH value. As the pH value drops in the milk, the positively charged hydrogen ions effectively neutralize the negatively charged "hairy" outer layer of the casein micelles. As a result, the micelles are suddenly unable to interact with the water molecules in solution and begin to separate from the water and coagulate together instead. Eventually, the coagulated micelles form a three-dimensional web-type structure that captures the water and other components in the milk, primarily lactose, fat, whey proteins, and minerals. The casein structure is rather weak and incapable of contracting to expel the water and other compounds. As a result, acid-coagulated cheeses are very high in moisture content, which renders them susceptible to microbial spoilage. They are soft and generally meant to be consumed fresh. Common examples include cottage cheese, cream cheese, and the aforementioned *quark* (Donnelly 2014).

Acid/heat coagulation makes use of—wait for it—acid and heat to initiate coagulation. When milk that has been slightly acidified either through the action of lactic acid bacteria or by the addition of an acid like vinegar, it will coagulate at elevated temperatures (85°C / 185°American). The whey proteins, which are not very heat stable, unfold and become less soluble in water. The "hairy," negatively charged surfaces of the casein micelles are neutralized, allowing the unfolded whey proteins to bond to the micellar surfaces. The micelles then aggregate into a network of curds that also traps the fat particles. The curds float to the surface of the solution and can be separated from the remaining whey and allowed to drain further. These acid/heat coagulated products are also very high in moisture content and, therefore, subject to microbial spoilage and meant to be consumed fresh. The most common example of an acid/heat coagulated product is ricotta cheese. Another common example in which the curds have been pressed into a solid mass is queso blanco (Donnelly 2014).

The final and most common means of coagulating milk for cheese involves the use of rennet, a protein-degrading enzyme complex made up primarily of the enzyme chymosin as well as some pepsin. Traditionally, rennet has been derived from a ruminant's abomasum (the technical term for a ruminant's fourth stomach), the most common being calf rennet, but there is also goat kid or lamb rennet. As described in the previous chapter, this is how cheesemaking was stumbled upon in the first place, and technically speaking, an abomasum is the only source of true rennet. More modern sources of rennet include primarily plant-derived and bacterial- or fungal-derived enzymes. However, these are not rennet per se but rather coagulating enzymes or rennet substitutes. But we're not picking nits here and will refer to all sources of the coagulating enzymes as rennet, regardless of their actual sources. Interesting, though, is the fact that any of the rennet not derived from animal sources is typically referred to as "vegetable" rennet, even if it's derived from bacteria or fungi. Even more interesting is that many of these rennet products are derived from genetically modified organisms (GMOs) that have had the gene for bovine chymosin spliced into the DNA of the host microbe, typically a yeast or other fungus, and are then grown in large vats so that the chymosin can be harvested (Donnelly 2014). I refer back to my previous statement about not picking nits, though.

The rennet enzymes effectively give the casein micelles a haircut by cleaving the charged κ-casein portion of the macromolecules. This exposes the hydrophobic (unable to mix with water) inner portion of the molecule and causes the micelles to aggregate together. As coagulation progresses, the casein aggregates form a large, three-dimensional netlike framework that also traps lactose, fat, whey proteins, and minerals. The difference between this network, however, and the one formed from acid coagulation is that is has much greater structural integrity and can better withstand more extreme methods of contraction to expel whey, such as cutting, pressing, and salting. The lower moisture content that results from removing the whey allows for cheeses that can be aged without fear of microbial spoilage. A second difference is that because acid is not used to do the coagulating, which also occurs much more rapidly with rennet and doesn't allow the lactic acid bacteria much time to produce acid and drop the pH, a wider range of starting pH values is available to the cheesemaker, depending on the desired style. This opens the door to many different ripening and aging options and is the reason why rennet coagulation is by far the most commonly used method for most types of cheese produced (Donnelly 2014).

Rennet-coagulated cheeses provide many more options for cheese production; but this comes at the expense of increased complexity in the production process. Whereas acid-coagulated cheeses are generally quick and easy to produce, they are also very limited as far as the styles of cheese that can be produced with this method. The fact that rennet-coagulated cheeses can be ripened for months or even years necessitates getting the conditions just right in the beginning so that the intended cheese going in is what you get at the end of the process. The three main factors that must be considered are the pH, the moisture content, and the salt concentration added. The desired pH range is typically in the 4.6–5.4 range, with moisture content in the 30–60% range and salt concentration in the 0.5–4.0% range, all depending on the style of cheese being produced. The challenging part is getting them all in the correct ranges for the desired style, as they all affect each other and the microbial and chemical processes that occur during ripening, not to mention the environmental conditions that play a role as well. Beyond these basic factors, however, there are eight steps involved in making rennet-coagulated cheese, namely, setting, cutting, cooking, draining, knitting, pressing, salting, and finishing.

Setting

Setting is the process wherein the milk is inoculated with the lactic acid bacteria starter culture and allowed a brief period for the bacteria to proliferate, followed by the addition of rennet and the resulting coagulation. Traditionally, milk was not intentionally inoculated with starter cultures. Instead, bacteria that found their way into the milk would produce similar, although inconsistent, results. Fortunately, the bacteria that grew under different conditions in different types of cheeses were eventually isolated and characterized for flavor production, heat and salt tolerance, and so on, and can now be purchased as starter cultures for different cheeses. The lactic acid bacteria used to inoculate milk for cheesemaking can be placed into two categories based on heat tolerance. The mesophiles that can withstand moderate temperatures are *Lactococcus lactis* spp. *lactis* and *Lactococcus lactis* spp. *cremoris*. The thermophiles that like it hot are *Lactobacillus delbrueckii* spp. *bulgaricus, Lactobacillus helveticus, Lactobacillus delbrueckii* spp. *lactis*, and *Streptococcus thermophilus*. The mesophiles can withstand temperatures only up to about 40°C (104°American), whereas the thermophiles can withstand temperatures up to about 65°C (149°American). The bacterial culture is allowed to grow and acidify the milk to the degree specified by the style of cheese being produced prior to the addition of the rennet. Coagulation generally occurs within 1 hour of adding the rennet. The firmness of the curd depends on the pH of the milk and the time elapsed following the addition of the rennet. The lower the pH and the longer the time, the firmer the curd will be. In what may seem somewhat paradoxical, a firmer curd will lead to a cheese with a higher moisture content, as a firm curd is less able to contract and expel whey during pressing.

Cutting

After coagulating, the curd is ready for cutting. This step is also critical in ultimately determining the final moisture content of the cheese. A smaller curd has a greater surface area to volume ratio and is better able to expel the whey. So, for a firmer cheese, such as a Parmesan style, the curd should be cut as small as possible, whereas for a softer style of cheese, such as a bloomy rind cheese, the curds can be cut into much larger pieces.

Cooking

The cooking step is a bit of a misnomer because it's actually cooking and stirring, but I suppose a suitable contraction (cookirring?) isn't really available. This is the step where heat is added to the cut curd while it is being stirred. Higher cooking temperatures with more stirring lead to greater contraction of the curds and more whey expulsion. That said, the lactic acid bacteria in the curd are still producing acid, which impacts the process as well, since a lower pH leads to greater contraction and expulsion of whey as well. This is where it gets a little more complicated, though, since the bacteria are more or less active depending on the cooking temperature. The cheesemaker must therefore balance the temperature with the resultant acid production to achieve the desired level of curd contraction and whey expulsion.

Draining

Draining, or what was traditionally referred to as dipping, is the stage when the liquid whey is separated from the solid curds. The traditional method of dipping involved scooping out the curds and whey into a perforated container to drain the whey. Tired of losing their curd scoopers and afraid that the term "dipping" would no longer apply, cheesemakers developed the modern innovation of a drain valve fitted with a screen on the same kettle used to coagulate and cook the curd so that they could simply open the valve and drain off the whey. And, hence, draining was born! In industrial settings, the entire mixture of curds and whey is often pumped into a separate vessel that's perforated or has the aforementioned drain valve.

Knitting

Knitting is the stage when the curds "knit" together into a gelatinous mass as the whey drains from the curd. Similar to the cooking step, temperature plays a critical role at this stage as well. Whereas higher temperatures generally lead to greater whey expulsion, the bacteria, which are still active, are also influenced by temperature. The bacterial activity, in turn, dictates how much acid is produced, which in turn influences the whey expulsion.

Pressing

Pressing, the next step, could certainly be lumped together with knitting, but I'll keep them separate for consistency's sake. This may be surprising information, but the cheese is pressed at this stage to further expel the whey and knit the curds together. Some cheeses are unpressed, which means that the whey is allowed simply to continue to drain via gravity; however, most cheeses are pressed. The process used to be rather crude, but fortunately more modern pressing devices have been developed to apply specific amounts of pressure depending on the style of cheese being produced. Pressing allows for the production of very closed textures and tight rinds, which are critical to the development of eyes (from the gases produced by bacteria and not of the seeing variety) and hard rinds, respectively.

Salting

In our never-ending quest to expel whey, salting is the next step. The addition of salt creates an osmotic gradient that draws additional whey from the curd. There are a few different methods for applying the salt. The simplest method is to rub dry salt onto the cheese. As the salt dissolves, it diffuses into the interior. However, as salt is continually applied to the exterior of the cheese, moisture is drawn out of the cheese. The moisture evaporates and this ultimately produces a hard rind, which helps to protect the cheese; but the formation of a rind can be problematic if more salt is to be added, as it becomes increasingly difficult for it to be incorporated into a cheese with a hard rind. A second method for adding salt is by using a liquid brine. This method is effective for making larger rind cheeses, as the brine allows for greater salt absorption while more gradually dehydrating the exterior of the cheese. The final method for adding salt is to mix it directly with the curd particles prior to pressing. This affords a much more even distribution of the salt throughout the cheese, without the distinct gradients of salt that results from the former two methods. The disadvantage of this method is that it is very difficult to develop a rind this way, as the exterior is not dehydrated like with the other methods.

Finishing

The final step in rennet-coagulated cheese making is, well, finishing. As the name suggests, this is really a ripening stage when the curds are converted from semisolid nuggets that taste pretty much like raw milk to a firmer or creamier textured cheese with a mature flavor profile. The process can be quite distinct for the interior of the cheese as compared to the surface. However, both (interior and exterior) are dependent on the microbes introduced or indigenous (if you're old-school or unsanitary); environmental conditions such as temperature, oxygen exposure, and humidity; and any physical manipulations like washing and turning. The microbes respond, in turn, to the environmental conditions and physical ministrations to produce flavor and textural changes resulting from the breakdown of proteins and fats in the ripening cheese. The entire ripening process can take as little as a few days to as long as a few years for some cheeses (Kindstedt 2013).

Overall, dairy fermentation is a fascinating process that began as a necessary invention to preserve a highly nutritious, yet highly perishable, food source and has been used by humans for thousands of years to produce a variety of different food products with unique flavors and textures. By understanding the science behind dairy fermentation, we can better appreciate the delicious foods that it produces and continue to innovate new dairy products in the future.

Chapter 13

PRESENT

—

The current landscape of fermented dairy is very different from the historical ferments that were born of necessity. For example, there are thousands of named cheese varieties today. That may seem a bit overwhelming. Fortunately, the vast majority of these cheeses belong to one of a few main cheese families that share similar manufacturing protocols and compositional characteristics, which certainly helps when describing modern cheese production. Those families are fresh, bloomy rind, smear ripened, hard uncooked, hard cooked, and blue (Donnelly 2014).

Modern cheese production in the South has certainly come a long way from its humble beginnings. Nowadays, it has a rich landscape of flavors and textures that rivals some of the traditional cheesemaking capitals of the world. Let's take a look at a few of the standouts, in my view. Mind you, this coverage is by no means even remotely complete. It's merely a glimpse at some of the producers with whom I happen to be familiar. I encourage you to get out there and sample as many different options as you can!

ASHE COUNTY CHEESE

I'll start with Ashe County Cheese because they're practically in my backyard—just down the road from Boone in West Jefferson, North Carolina, a sleepy little mountain town that has become quite the tourist destination in the last several years—and because one of the current owners, Lucas "Luke"

Everhart, was a student of mine. Ashe County Cheese began as a Kraft facility in 1930. Kraft corporation worked to consolidate several small, local producers into the one facility that produced cheddar for national distribution. Kraft operated the plant until 1975, when it was sold to the manager at the time, Chesley Hazlewood. After Mr. Hazlewood's passing in 1980, Mrs. Hazlewood sold the company to a couple guys from Wisconsin, Jerry Glick and Doug Rudersdorf, who upgraded the facility and store and made Ashe County Cheese the popular tourist attraction that it still is today. After a couple more ownership changes, Ashe County Cheese was bought by Newburg Corners Cheese, Inc., owned by Mike Everhart and Tom Torkelson, who with their families still own it today. So what makes Ashe County Cheese special? It is the oldest, continuously operating cheesemaking facility in North Carolina. Besides that, they have an impressive array of products that can be purchased from their store, ranging from a dazzling display of cheeses (including hoop cheese and cheese curds because, well, Wisconsin) to a host of other products, including salsas, fudge, and jams. You can also watch the cheese being made through a viewing room in the cheese plant. They have a new food truck as well, at which you can fill your face with all kinds of cheesy goodness. Not to mention that West Jefferson is a jewel of a town in the southern Appalachians that's worth the visit on its own.

BOXCARR HANDMADE CHEESE

From a historical and more industrial cheesemaking facility in the western part of North Carolina, we now travel to a smaller, artisanal cheesemaker in the Triangle area of the state (Raleigh, Durham, and Chapel Hill). Boxcarr Handmade Cheese, in Cedar Grove, North Carolina, is a family owned cheesemaking creamery established in 2015 and fashioned after small Italian farmstead creameries. I confess that while it is truly one of my current favorite cheesemakers, I also want to highlight them because another of our former Fermentation Sciences students is part of the family and worked at the facility for some time.

Siblings Samantha and Austin Genke are the owners, and Samantha is the head cheesemaker, while Austin designs and builds everything needed in the facility. The creamery complements the family farm, Boxcarr Farms, where they raise dairy goats, heritage pigs, chickens, and turkeys, while also growing heirloom produce. In the farm's creamery, they use the milk from their goat herd as well as from small, local, family-owned dairies. Besides making use of family members, they also provide cheesemaking internships and employ youth through the Orange County Second Family Foundation, a nonprofit that helps teens who are experiencing risky situations, such as foster care. So they're a pretty awesome small business. But, like I mentioned before, they also make a killer lineup of cheeses. They specialize in bloomy rind cheeses, which are some of my favorites. I think I might sell my liver for their Rocket's Robiola, an award-winning cheese they describe as "somewhere between a vegetable-ashed, Loire Valley, bloomy rind and a Piedmontese Robiola" that is treated with *Geotrichum candidum*, a mold that produces the bloomy, white, wrinkly rind and a lusciously creamy interior and is dusted in vegetable ash to finish it off. They make cheeses using cow and goat milks individually and also in combination. They also make a couple beer-washed cheeses that, sadly, I have not had the opportunity to try. So, if you're reading this, Samantha and Austin, feel free to send me some anytime you'd like.

SEQUATCHIE COVE CREAMERY

Despite what you may be thinking at this point, I do not exclusively eat cheese from North Carolina. I am as inclusive as it gets when it comes to cheese consumption, an equal opportunity cheese cater, if you will. Sequatchie Cove Creamery on Sequatchie Cove Farm in Sequatchie, Tennessee (sorry, I just really like that name, Sequatchie), which is just northwest of Chattanooga, produces European-style cheeses with a southern flair. Nathan and Padgett Arnold have been making award-winning cheeses there since 2010 using milk from the pasture-raised cows on their farm. They currently make four different cheeses, Cumberland, Walden, Coppinger, and Shakerag. Cumberland, named after the Cumberland Plateau, a densely forested area with rock outcroppings and caves where the farm and creamery are located, is an Alpine style Tomme with tasting notes of buttery potato and fresh corn; it is as beautiful as the surrounding region that inspired it. Walden, which is named for Walden Ridge, overlooking the Tennessee River, is a washed-rind, creamy cheese based on the Reblochon style of France that is reminiscent of walnuts, cultured butter, and sautéed mushrooms. Coppinger is the refined and elegant member of the lineup. Named after Coppinger Cove, where the creamery is located, this is an ash-lined cheese in the style of a Morbier, which gives it a distinct bifurcated appearance, with flavors of sweet cream, toasted almonds, and rising bread. Finally, the bad boy of the bunch, Shakerag, named after Shakerag Hollow, where moonshine used to rule the day, is the heavy metal complement to Coppinger's classical leanings. This is a blue, wrapped in fig leaves that have been soaked in Tennessee whiskey, and that tastes of candied bacon, toasted coconut, and cacao nibs. If this doesn't make you want to jump off your sofa and go buy a wheel, you may want to check your wrist and make sure you still have a pulse. In fact, I may have to go take a small Uncle Nearest (a delicious Tennessee whiskey that is described in chapter 18) and Shakerag break.

SWEET GRASS DAIRY

Our next amazing southern creamery is Sweet Grass Dairy of Thomasville, Georgia, which is just over the border from Florida, about 45 minutes north of Tallahassee. Al and Desiree Wehner started Sweet Grass Dairy in 2000 after changing their dairy farm from conventional farming methods to an intensive grazing program about a decade earlier. Since then, the cows have been strictly pasture raised on the abundant, rich grasses of southern Georgia. Desiree wanted to show off the delicious cheese that is made with milk from cows raised purely on grass grown under the warm Georgia sun. As the business grew, their daughter, Jessica, and son-in-law, Jeremy Little, joined the business and then eventually took it over in 2005. From there, the Littles have never looked back and now have a restaurant in downtown Thomasville, where their award-winning cheeses, along with other specialty items, are sold and a full food and drinks menu is served.

Their cheese selection is certainly impressive, with several aged cheeses as well as a few spreadable cheeses, including the obligatory pimento cheese. Their pimento cheese is unique, however, in that they use their Thomasville Tomme as the base. The Tomme is based on the classic French Pyrenees Tomme and is made with raw cow's milk, as are their other aged cheeses. Green Hill is named after the Wehner's first rotational-grazing dairy and is a double cream, soft-ripened cheese with a creamy texture and subtle mushroom flavor reminiscent of a Camembert. Asher Blue, named after the Little's second son, is a creamy blue cheese with a natural rind and an earthy flavor with a hint of cocoa on the back end. Their Georgia Gouda is a young, tangy Gouda with a sweet, buttery finish. And last,

but certainly not least, is Griffin, named for son number three, which is a farm-style cheese that has been soaked in Terminus Porter from Gate City Brewing and aged for a minimum of 120 days.

MEADOW CREEK DAIRY

Meadow Creek Dairy in Galax, Virginia, just over the border from North Carolina and in the beautiful southern Appalachian Mountains, is a farmstead dairy based on the long and storied tradition of Alpine cheesemakers. The cheesemakers' herd of cows, with breeds that hail from Alpine regions around the world, is purely grass fed, which shines through in their lineup of raw milk cheeses. Appalachian is their original recipe and was developed to highlight the flavors of the milk from cows that graze solely on the beautiful mountain pastures of southern Virginia. It has a bloomy rind of *Penicillium* with a soft interior and flavors of mushroom, cream, and butter. In case you want to go all out, though, they also offer the Extra-Aged Appalachian, which is aged for 8 months (as opposed to the 2 months the standard Appalachian is aged). This creates a completely different texture and flavor profile. The texture is firmer and the mushroom, earthy flavor is complemented by a fruity tanginess. Grayson, named after the county they are in, was inspired by trips to Wales and Ireland. It's a washed-rind cheese with a red-orange rind and a creamy, fudgy texture. The flavor is grassy and nutty like the finest examples of washed-rind cheeses. The Mountaineer is a classic Alpine-style cheese that showcases their own mountainous terroir. Aged for at least 6 months, this Alpine-style cheese is firm but creamy, with a roasted, nutty flavor and a hint of caramel. Finally, Galax is their Dutch-style Gouda named in honor of their hometown. It's a young Gouda, with a smooth, buttery texture and flavors of lemon, tangy yogurt, and walnuts.

I could keep going on about all the amazing cheesemakers in the South, but for the sake of brevity, I just wanted to highlight a few that I'm somewhat familiar with. I encourage the readers of this book to get out there and find some cheesemakers close to you. For those of us in North Carolina: Were you aware that there's even a Western North Carolina Cheese Trail? Neither was I, until recently. Trust me, I intend to hit the cheeses on the list that I haven't tried yet. And, finally, don't miss the annual Carolina Mountain Cheese Festival that benefits the Western North Carolina Cheese Trail.

OTHER PRODUCTS

You may be thinking that I have forgotten that there are fermented dairy products other than cheese. I have not. But, as far as I know, there really aren't any yogurt or kefir or other fermented dairy producers of note in the South. There may be some small, family farm producers, but the market seems to be dominated by rather large companies. This is particularly true with yogurt. Besides the mega companies like Dannon and Yoplait, the others on the shelves are pretty big, too, with Greek yogurts like Fage, Oikos, and Chobani (which used to have their sales headquarters in Charlotte, which is something) dominating the shelf space. I guess it makes sense. Whereas even a standard grocery store may carry dozens or sometimes even hundreds of brands of cheese, yogurts are much more limited. And kefirs even more so. I think it goes back to the fact that while we enjoy yogurt and kefir as healthful and tasty additions to our daily intake of calories, they are not sought-after specialty items like fine, artisanal cheese. I suppose it's the same reason why the beer and wine shelves are always stocked full of hundreds of different brands as well, while the juice aisle is generally far more limited. Well, perhaps the future of fermented dairy in the South will provide more options aside from cheese. Although, please keep the cheese coming!

Chapter 14

FUTURE

—

The future of fermented dairy holds so many possibilities. Genetic modification of bacteria or using existing microbes to produce antibacterial compounds that can make dairy products safer is already becoming popular in the industry. You can now purchase commercial bacterial cultures for making fermented dairy products that produce bacteriocins to prevent pathogenic bacteria from growing. Fermented dairy is one of the most highly regulated food industries in the country, and for good reason. Infected dairy products can cause serious harm and even fatalities. Advancements to make the industry safer are always welcome.

ALTERNATIVE DAIRY FERMENTS

If you peruse the dairy aisle of any modern grocery store, the cheese selection is generally pretty decent (and getting better all the time), and now you have ample choices when it comes to yogurt as well. You'll probably even see kefir at most large grocers these days. That said, there are so many dairy ferments from around the world that make use of milk from a variety of animals and that are not currently on store shelves in this country. Where are the *smetanas*, *quarks*, and *filmjölks*? Some stores have started carrying *skyr*, which is the Icelandic version of yogurt, a thicker, creamier yogurt than even Greek yogurt. It's what you might expect if yogurt and cream cheese fell in love and had delicious, creamy children. These alternative dairy ferments could be incorporated into the southern diet to make the experience even more varied, healthful, and flavorful. Not to mention that spreading our reliance on cow milk to other animal milks could be beneficial to the environment as well.

Similarly, if you're perusing the vegan food products at your local grocery store, there are many more vegan cheese options. Yes, I know, these are not dairy ferments, but it also feels weird putting them in the "Fruits and Vegetables" part. And since they're "cheese," as in they're intended to give you a similar flavor and texture experience to cheese, just without the actual dairy that goes into making normal cheese, I'll keep them in the "Dairy" part, thank you very much. There are plenty of sliced American "cheese" and "cheddar"-type options available, but as far as I know, there are not many "Camembert" or "Robiola" options out there yet. That is, the vegan options for the finer aged and bloomy rind cheeses are limited at best. This is a sector of the market that could see tremendous growth in the future, as there is very little in that sector currently. Animal milk has a very particular chemical makeup that we have taken advantage of to hone our cheesemaking skills over thousands of years. It's difficult to mimic this chemistry to produce a nondairy fermented product that is similar to the real thing. Creating unaged "cheeses" that mimic the texture and flavor of cheese is less of a challenge; but getting the creaminess of the interior of a well-aged Brie in a vegan product is far more challenging, which is why we have not seen many yet. The ones that are out there are generally made from cashews, which provide some of the creaminess, but I have yet to taste one that competes with the real thing.

UTILIZING WHEY

Figuring out what to do with all the whey, the major byproduct of cheese production, is an ongoing issue. There's only so much whey protein powder you can make. Sure, there are whey cheeses, like ricotta, but imagine if we made ricotta from the whey from every batch of cheese produced. Considering that you get an average of 9 pounds of whey for every pound of cheese produced, that would be a lotta ricotta! That's not even considering the whey from Greek yogurt production, in which the whey is strained out of the yogurt to create the thicker texture. Greek yogurt production results in roughly twice as much whey as yogurt produced.

For the sweet whey from cheese production, there are certainly applications, such as whey powder production, but the volumes required to make this a viable solution are generally far out of reach for the average artisan cheesemaker. Another option includes spreading it on agricultural fields as a fertilizer, but this is heavily regulated, as it can easily pollute waterways if too much is applied. Pigs and chickens love the stuff, so feeding it to them is a possibility, but these arrangements can be difficult to establish and can also be expensive, since many farmers won't pick up the whey unless they are paid to do so.

For the acid whey produced from Greek yogurt, there are even fewer options available. Biodigesters are a possibility for either type of whey. These are anaerobic digesters that use the whey to produce biogas, which can in turn be used to generate electricity. However, these require massive amounts of whey to make them even remotely viable as an energy source.

More recently, enterprising cheesemakers and scientists have teamed up to make some novel products from whey, such as beer and distilled neutral spirits. Whey has very little sugar, though, so when used to make a distilled spirit there is still a lot of waste at the end of the process. That said, the value addition of making vodka from whey, for example, which has been done by Vermont Spirits and Hartshorn Distillery in Tasmania, might be enough to pay for the disposal of the waste. Making beer, though, may be the more attractive option. Samuel Alcaine, an associate professor at Cornell, has experimented with this and has come up with a viable, low ABV (alcohol by volume) product from the whey. I'm not sure how this could technically be called a beer if there are no grains used, but the process uses all the whey, which is great. Alcaine has also suggested that acid whey could be used to replace the water in sour beer production, thus adding the acidity without any additional microbial action necessary. As they say, where there's a will (or a still!), there's a whey.

PRECISION FERMENTATION AND ANIMAL-FREE DAIRY

My final dairy prognostication involves other products that push the boundaries of what are considered dairy products. Precision fermentation, which was discussed in chapter 4, can be used to produce the chemical compounds that constitute milk, such as casein, the primary milk protein.

Once the individual components have been produced, they can be added together to make animal-free dairy products. These are really interesting, as they are effectively real dairy products, just produced with no cows involved. All the major components of milk are produced through microbial fermentations and then put together into a product that looks, smells, and tastes like the real thing. Alternatively, some of the animal-free dairy components can be combined with plant-based products like coconut milk to make hybrid products. Even the triacylglycerols that constitute roughly 98% of the fats in human milk are being produced by the oleaginous yeast *Yarrowia lipolytica* for inclusion in infant formulas now—to better mimic the real thing. Using precision fermentation to produce compounds such as this as well as lactoferrin, an iron-binding protein found in colostrum and breast milk, not only makes infant formula more nutritious but also helps to overcome some of the manufacturing and supply-chain issues that have plagued that industry in recent times.

Perfect Day, a precision fermentation company based out of Salt Lake, was the first such company in the United States to receive approval to produce animal-free milk proteins. They are still the biggest player in the business and supply their milk proteins to a number of companies making all sorts of animal-free dairy products, including ice cream, yogurt, and milk chocolate, among others. Even large, global companies like Nestlé, Unilever, and Starbucks are starting to make and use animal-free dairy products now. The beauty of animal-free dairy products is that they look, feel, and taste just like the real thing, but are made from individual components derived from plant-based and fermentation products that have the potential to reduce the environmental and energy footprints of the animal-derived products. So I say, let's milk this for all it's worth!

Chapter 15

RECIPES

—

I'm sticking to the basics here, while still having some fun. Cheesemaking takes a lot of practice, and it's good to hone your technique before jumping into more advanced, longer aged, and bloomy rind cheeses. If I were doing a book on dairy fermentations alone, I would start with the basics and slowly work up to the more advanced ferments. Since I'm doing a book on lots of different types of fermentations, it wouldn't make sense to dive into the deep end of dairy fermentations. So, let's look at some straightforward, but interesting and delicious, dairy ferments.

For cultures, there are a lot of great sources out there. I have used cultures from New England Cheesemaking Supply Co. (https://cheesemaking .com), Cultures for Health (https://culturesforhealth.com), and Yógourmet (https://yogourmet.com), all with good success. For a kombucha SCOBY for the wheybucha, White Labs (www.whitelabs.com) or Cultures for Health are the sources you need.

HOOP CHEESE

Start to finish: 18–24 hours | Yield: A little under 1 kg

This is about the most basic cultured cheese you can make. It's effectively an unaged cheddar with no salt added and without the time-consuming cheddaring process. It is consummately southern, though, and where it lacks a bit in flavor (no salt will do that), it melts easily and works great in mac and cheese or grilled cheese.

Ingredients

- 8 L whole milk
- 2 g Meso II mesophilic starter culture
- 2 g calcium chloride solution diluted in 60 mL filtered water
- 3 g liquid rennet diluted in 60 mL filtered water

Instructions

Heat the milk slowly in a pot set in a water bath at about 35°C (95°F) until the milk reaches 30°C (86°F).

Once the milk reaches temperature, turn the heat off in the water bath, and sprinkle the starter culture over the milk. Let the starter rehydrate for about 5 minutes and then gently mix it into the milk using a whisk.

Cover the milk and maintain the 30°C temperature for 1 hour to let the cultures start acidifying the milk.

After 1 hour, add the calcium chloride diluted in water and gently but thoroughly whisk it into the milk. Next, add the rennet diluted in water and whisk it gently but thoroughly into the milk as well.

Cover the milk again and let it sit for about 45 minutes at 30°C or until the curds give a clean break (when you can gently cut into the solid curd with a knife and the whey that fills the gap is clear).

At this point, cut the curds into 1.5 cm pieces and let them rest for 5 minutes. After 5 minutes, slowly increase the heat to 39°C (102°F) over about 40 minutes.

Turn the heat off after the milk reaches 39°C and, maintaining this temperature, gently stir the curds for about 20 minutes, or until they are about the size of peanuts and have firmed up.

Now, let the curds rest at this temperature for another 30 minutes. They should sink to the bottom of the pot at this point.

After 30 minutes ladle off enough whey to expose the tops of the curds. Now, stir those curds for about 15 minutes until they start to mat together. They should stick together when pressed in your hand.

Ladle the curds (a skimmer works great for this) into a colander lined with cheese cloth or butter muslin and let them drain for about 5 minutes.

Line a 20 cm Tomme mold with damp cheesecloth, ensuring that there are as few folds and wrinkles in the cheesecloth as possible. Gently transfer the curds to the mold and fold the remaining cheese-cloth over the top of the curds, again ensuring there are as few wrinkles or folds in the cloth as possible.

Set a follower (a lid for the cheese mold) on top of the mold and press at 8 pounds for 1 hour, then remove the cheese from the mold, unwrap it from the cheesecloth, flip it, put it back in the cheesecloth (or not, if it's holding together enough already), and press it at 10 pounds for about 12 hours.

At this point, you can remove the cheese from the mold and enjoy. You can also wax the cheese and age it for a bit in a cool, humid environment (10°C / 50°F and 80% humidity), but with no salt and a high moisture content, it's not really made to age and may spoil if aged for too long.

Fermentation

The Meso II mesophilic starter culture is a long name for a culture that contains only one organism, *Lactococcus lactis* ssp. *cremoris*, a very common lactic acid bacteria used in the production of all kinds of fermented dairy products. The mesophilic in the name refers to the fact that these bacteria are more active in moderate temperature ranges, which also makes them easy to use for fermentation. They consume the lactose in milk and convert it to lactic acid, thereby acidifying the milk and beginning the coagulation process (which will be completed through the addition of the rennet). However, these bacteria also appear to have some pretty significant health benefits as well. Using *Drosophila* (fruit flies) as well as rodent models, a recent study showed that *Lactococcus lactis* ssp. *cremoris* bacteria provided anti-inflammatory activity in the gut as well as overall protection against tissue damage in the intestines (Darby et al. 2019). Another study using human models this time showed that daily intake of the bacteria reduced morning cortisol levels in the test subjects, indicating a reduction in psychological stress levels (Matsuura, Motoshima and Uchida 2022). So, not like you needed extra motivation to eat cheese on a daily basis, but just in case you did, there you go.

....

YOGURT

Prep time: 1–2 hours | Fermentation time: 6–12 hours | Yield: 1–2 L

Ingredients

- 1–2 L milk (I use whole milk that is pasteurized and homogenized, but lower fat milks work as well. You just won't get as creamy a finished product as with whole milk.)
- Yogurt starter culture (some are made for use with 1 L of milk, and some are made for half gallons of milk, which is close enough to 2 L)

Instructions

Heat the milk slowly to 82°C (180°F) or up to a boil (this is unnecessary, but if you're using the yogurt-maker function on your pressure
cooker, this may be the only option). A double boiler works great for
this step. If you have control over the process, holding the milk at
82°C (180°F) for 10 minutes will result in a thinner yogurt, whereas
holding it for 20 minutes at that same temperature will result in a
thicker yogurt.

Cool the milk down to 42–44°C (108–111°F) and pour it into one or
more jars (or not, if it's already in your yogurt maker).

Pitch your starter culture by either sprinkling it on top of the cooled
milk, allowing it to sit for 2 minutes, and then gently stirring it in, or
by doing the same in a separate cup (accuracy is not paramount to
this volume) of the cooled milk and then incorporating it into the rest
of the milk.

Place a lid on the jar with a fermentation cap or just leave the lid
on the jar loose to allow the gas to escape. You can also put a clean
towel secured with a rubber band over the top of the jar.

Ferment the yogurt at 42–44°C (108–111°F) for 6–8 hours or until the
yogurt has reached the desired level of tartness and thickness. The
longer you let it go, the firmer and more sour it will become. Don't
fear if you let it go for more than 8 hours. It can easily go for more
than 12 hours with no problems.

After fermentation is complete, place the yogurt in the refrigerator
for at least 8 hours.

At this point, you can enjoy the yogurt. There will likely be a bit of
whey that has separated. You can either pour it off carefully or you
can strain the yogurt through cheese cloth for a thicker product.
Either way, collect your whey to make some wheybucha (the recipe
is below!).

Fermentation

Heating the milk to 82°C (180°F) not only kills any bacteria that may already be present in the milk, but it also denatures the whey proteins (that is, disrupts the molecular conformation of the protein), leading to an increase in viscosity and thicker overall texture in the final product. As noted above, if you hold that temperature for a little longer, you end up with a thicker yogurt. After cooling the milk to 42–44°C (108–111°F), you add your starter culture, which will generally be a mix of several different thermophilic (heat loving) lactic acid bacteria. The essential bacteria are *Streptococcus thermophilus* and *Lactobacillus delbrueckii* ssp. *bulgaricus*, but some other common ones include *Bifidobacterium lactis*, *Lactobacillus acidophilus*, and *Lactobacillus delbrueckii* ssp. *lactis*. Depending on the culture you use, the mix of bacteria may be a subset of these or have even more bacteria. As the bacteria consume the milk sugar lactose, they produce lactic acid, thereby acidifying the yogurt and lowering the pH. Prior to acidification, the small particles (referred to as micelles) of casein protein, the primary milk protein, have a negatively charged outer layer. This prevents the particles from coagulating, as like charges repel each other. Once the pH drops to 4.6 or below, those casein particles are no longer charged, and the micelles coagulate, thickening the yogurt. The denatured whey proteins from the heating phase can also join in the coagulation party. That's why a longer heating time leads to a thicker yogurt. More whey protein is denatured and adds to the coagulation of the casein micelles, creating a thick, viscous yogurt. In general, the higher the protein content in the milk, the thicker the resulting yogurt will be.

However, in addition to the total protein concentration, other factors play a role in the texture of the finished yogurt. A lower fermentation temperature (below 40°C / 104°F and down to as low as 30°C / 86°F) actually leads to a smoother, creamier mouthfeel in the yogurt. As the proteins coagulate, they form a mesh-like three-dimensional network. Lower temperatures during this process produce a more stable three-dimensional network that ends up expelling less whey. As a result, you get a yogurt that is smooth and creamy with less whey separating from the yogurt. The only issue with this method is that the time that it takes for the yogurt to coagulate is greatly increased (like up to 18 hours or more sometimes). If you're patient, though, you may be rewarded with extra creamy yogurt.

KEFIR

Prep time: 1 hour | **Fermentation time:** 24 hours
Chilling time: 8 hours | **Yield:** 1–2 L

Ingredients

- 1–2 L milk (depending on the volume of milk specified for the starter culture)
- Kefir starter culture

Instructions

Heat the milk slowly to 82°C (180°F) or up to a boil (still unnecessary, but no worries if you forget about your milk for a bit and it gets to this point).

Cool the milk down to 23–25°C (73–77°F) and pour it into one or more jars.

Pitch your starter culture by either sprinkling it on top of the cooled milk, allowing it to sit for 2 minutes, and then gently stirring it in, or by doing the same in a separate cup (accuracy is not paramount to this volume) of the cooled milk and then incorporating it into the rest of the milk.

Place a clean towel over the jar and secure it with a rubber band.

Ferment the kefir at room temperature until a curd forms (this should take about 24 hours).

Refrigerate the kefir for at least 8 hours and then stir it to break up the curd and liquefy it.

At this point, you can enjoy it straight, or it goes great with fruit or other flavorings. Blueberry kefir and lemon kefir are 2 of my favorites. For the lemon kefir, just zest some lemon into the kefir. You can add some lemon juice too, but this may make it overly acidic.

Fermentation

Kefir originated in the Caucasus Mountain region of Russia. It is one of the most consumed cultured dairy products in the Middle East, Eastern Europe, and Central Asia. I've even heard it referred to as the champagne of cultured dairy since it has some effervescence from the carbon dioxide production. That may be a bit of a stretch, but it is delicious and very good for you. The cultures used to produce kefir are a mix of lactic acid bacteria, yeast, and sometimes even acetic acid bacteria. Every culture is different, but some of the more common bacteria include *Lactococcus lactis* ssp. *lactis* and *cremoris*, *Lactococcus lactis* ssp. *lactis* biovar *diacetylactis*, *Leuconostoc lactis*, *Leuconostoc cremoris*, *Lactobacillus brevis*, *Lactobacillus kefir*, and sometimes also *Lactobacillus delbrueckii* ssp. *bulgaricus* and *Lactobacillus acidophilus*. The yeast is generally either *Candida*, or *Kluyveromyces*, or *Saccharomyces* species, or some combination thereof. The mix of homo-fermentative and heterofermentative lactic acid bacteria produces a com-bination of lactic and acetic acids, diacetyl (butter flavor), and acetaldehyde from the lactose (acetaldehyde is often described as green apple flavor, but to me it's more reminiscent of bruised red apple or pumpkin guts). The pro-duction of these flavor compounds give kefir a more complex flavor profile than its yogurt cousin. The yeast also produces some alcohol, so that some products can contain up to 2% by volume (this is controlled in this country, but the traditionally made product often has some alcohol), in addition to carbon dioxide, giving the kefir its effervescence. The possible presence of acetic acid bacteria can sometimes, if exposed to oxygen, convert the alco-hol to acetic acid in much higher concentrations than that produced from the heterofermentative lactic acid bacteria, giving the kefir an unpleasant vinegar flavor.

The fermentation often results in the production of kefir grains as well. These are composed of coagulated proteins, carbohydrates, and microbes (often where the acetic acid bacteria reside). In traditionally made kefir, these would be sieved from the product and used to inoculate subsequent batches. They can still be used for that purpose, but with such a mixed-culture fermentation, certain microbes will end up dominating after a couple generations and the final product may be vastly different from the original batch.

....

SKYR

Prep time: 1 hour | Fermentation time: 36–72 hours | Yield: 1 L

Skyr is the Icelandic version of yogurt, even though it's technically a cheese. If you like the thick, creamy texture of Greek yogurt and you haven't yet tried skyr, then you really need to. It makes Greek yogurt look thin and runny in comparison. The production process is very similar to that of yogurt, but with one additional step and a little rennet to really get a clean curd break and a nice, thick finished product. I love it and hope you will, too.

Ingredients

- 2 L + 250 mL milk (traditionally skim milk, but it can be made with whole milk as well)
- 0.4 g skyr starter culture (1 packet)
- ¼ rennet tablet (or 1 g of liquid rennet)

Instructions

This is basically a 2-step process. First, you are going to make a small batch of yogurt, and then you will add that as a starter culture to make your skyr.

Pour 250 mL of cold milk into a small glass jar. Add the starter culture and mix well.

Place a small towel or coffee filter on the top of the jar and secure it with a rubber band.

Allow the milk to ferment at warm room temperature (21–25°C / 70–77°F) for about 12 hours or until it has set. This may take quite a bit longer than 12 hours. Don't be surprised if it takes about 48 hours. This is fine.

Once it has set, cover the jar with a lid and refrigerate it for at least 8 hours.

Now, heat 2 L of milk slowly to 82°C (180°F) and hold it for 20 minutes at this temperature.

Cool the milk to 32°C (90°F).

When the milk is almost to temperature, mix the ¼ rennet tablet into 60 g (60 mL) of filtered water and stir to dissolve. Make sure to use this rennet solution within 5 minutes of preparing it. If you are using liquid rennet, measure 1 g of the rennet and dilute it in 20 g (20 mL) of filtered water at this point.

When the milk is at temperature, pour the rennet solution in and mix it thoroughly for no more than 15 seconds using an up and down motion.

Gently pour the milk into a jar or other container and mix in the yogurt starter culture that you prepared earlier.

Cover the container with a towel or coffee filter and secure it with a rubber band.

Ferment the skyr at warm room temperature (21–25°C / 70–77°F) for 12–18 hours or until it has fully set. You will know it's ready when the skyr has coagulated and the curd mass is floating above the separated whey (it will make a lot of whey).

When it is done, cover the container with a lid and refrigerate it for at least 8 hours. Now you can strain off the whey simply by pouring it out, or if you want a thicker skyr, you can strain the whole thing through cheesecloth (remember to collect your whey to make wheybucha—see the next recipe!).

Enjoy your delicious Icelandic yogurt/cheese!

Fermentation

Despite the fact that it's technically a cheese, as rennet is added to aid coagulation, skyr contains many of the same bacteria as yogurt, and most cultures will almost certainly contain *Streptococcus thermophilus* and *Lactobacillus delbrueckii* ssp. *bulgaricus*, the essential bacteria of yogurt. In addition to these thermophilic bacteria, however, skyr cultures may also contain some mesophilic bacteria that like it more room temperature rather than warm. The thickness of skyr is the result of several factors. For one, the addition of rennet greatly aids in the coagulation of the milk proteins. Note also that the milk is inoculated at a warmer temperature (32°C) and then allowed to cool to room temperature. As described above, fermenting at a cooler temperature helps to stabilize the coagulated protein network, making a creamier, thicker product. That said, the addition of rennet certainly ends up expelling a good bit of whey, unlike with yogurt fermented at cooler temperatures. The cooler temperatures also activate any mesophilic bacteria in the culture. Finally, skyr really should be strained through a cheesecloth after refrigerating it to make the truly thick, traditional product. Once you try it, you won't be disappointed.

....

WHEYBUCHA

Prep time: 15 minutes | Fermentation time: 1–2 weeks | Yield: 2 L

OK, now that we've made lots of whey, let's use it to make some kombucha, or wheybucha. The acid whey from the yogurt and skyr (although skyr is not as acidic as most yogurt) is arguably a bit better for this, as the lower pH adds a bit of extra microbial protection to the whey. If you're using the whey from cheese production, make sure that it's not a bloomy rind cheese or another cheese to which you've added mold cultures prior to collecting the whey. Otherwise, you will likely end up with some mold growing on top of your wheybucha. Trust me on this one. I speak from experience (oops!).

Ingredients

- 2 L collected whey
- 150 g sugar
- Kombucha starter SCOBY, or about 60 mL of fresh, live culture kombucha

Instructions

Pour the whey into a glass jar or other similar container.

Stir the sugar into the whey to dissolve it as well as possible. It's OK if you can't get it to dissolve completely at first. It will dissolve over time and feed the cultures. That said, dissolve it as well as you can initially.

Add the starter SCOBY or dose of fresh, live culture kombucha to the whey.

Cover the jar with a towel or coffee filter and secure with a rubber band.

Let the wheybucha ferment at room temperature for at least 1 week, or up to about 2 weeks. Check the pH after the second day to ensure that it is below 4.6. The pH should drop below 4 to about 3.5 or so by the time it is finished. After 1 week, taste it to see if it is acidic enough for you and ready to drink. The longer you let it go, the more acidic it will become, so if you want a wheybucha with a little more sweetness, then you probably want to stop it after 1 week or so.

Once it's done to your liking, cover the jar with a lid and put it in the refrigerator. It should last for a couple of weeks in the refrigerator. For the more culinarily adventurous, you can enjoy it straight. However, it also makes a great replacement for acidic dairy ingredients like buttermilk and yogurt in recipes and can be used as a base for soups as well.

Fermentation

Any kombucha fermentation involves a mixed culture of bacteria and yeast. In fact, SCOBY is an acronym for Symbiotic Culture Of Bacteria and Yeast. There are generally a number of different bacteria, often including *Bacillus* sp., *Acetobacter tropicalis*, *Gluconacetobacter saccharivorans*, *Micrococcus* sp., *Gluconacetobacter rhaeticus*, and *Paenibacillus taichungensis*, many of which are acetic acid bacteria, aerobic bacteria that convert alcohol to acetic acid. A number of different yeast species are found in a kombucha culture as well, often including *Brettanomyces bruxellensis*, *Saccharomyces cerevisiae*, *Zygosaccharomyces* sp., and *Candida* sp., or other similar ones. The yeast and bacteria work together to make the final product. The yeasts in the culture convert the sugar in solution to alcohol, which is then converted to acetic acid by the acetic acid bacteria. That's why the longer you let a kombucha ferment, the more acidic and vinegary it becomes. Since the bacteria are aerobic, they need access to oxygen. So, they excrete extracellular polysaccharides (pretty much cellulose), which forms a raft that they can live on and by means of which they can access oxygen while also fermenting the solution below. Pretty nifty. The raft of course is the SCOBY.

Now, go make some wheybucha and save that SCOBY in a little filtered water or diluted kombucha; then use it to inoculate your next batch. You can feed it with a little sugar to keep it viable longer, but make sure it has access to oxygen or the bacteria will eventually die off. Or dry your SCOBY for some vegan leather.

....

CRÈME FRAÎCHE

Prep time: 20 minutes | Fermentation time: 12–48 hours
Chilling time: 8 hours | Yield: 1 L

Crème fraîche is pretty much like fancy sour cream. It's a little thicker and not quite as sour, which makes it generally more palatable than its sour cousin. Even so, it can be used as a substitute for sour cream in any recipe or on baked taters. Unlike sour cream, it is also great with fruit-based desserts (or straight fruit) as an ice cream or whipped cream replacement. It is ideal for making cream sauces when stirred in toward the end. It doesn't curdle or "break" when heated, so it's a better option than most other dairy products for this purpose. It's also really easy to make, so let's get started.

Ingredients

- 1 L heavy cream
- 2 g Mesophile Aroma B starter culture

Instructions

Pour the cream into one or more glass jars (they can be pretty full) and place the jars in a pot large enough to fit them.

Fill the pot with enough water so that the water is halfway up the sides of the jars.

Over low heat, slowly heat up the cream until it reaches 30°C (86°F). Cut the heat off when it is still a couple degrees cooler than that so that you do not overshoot the temperature.

Keep the cream in the pot with the heat off to maintain the 30°C temperature (another reason to turn the heat off before reaching the target temperature). Sprinkle the starter culture over the cream and let it rehydrate for about 5 minutes. After 5 minutes, whisk the starter culture into the cream using an up and down motion.

Place the lids loosely on the jars and either leave the jars in the pot with warm water and move the whole setup into a cooler to maintain the temperature, or use a fermentation cabinet or other means to maintain the temperature for about 12 hours.

After 12 hours, take the jars out of their warm environment, tighten the lids, and let them continue fermenting at room temperature for a minimum of 6 hours or up to another 36 hours. The longer you let them go, the thicker and slightly more acidic the final product will be.

Refrigerate the crème fraîche for a minimum of 8 hours and up to 2 weeks before enjoying. This will further thicken it and allow the flavor to develop.

Enjoy!

Fermentation

The Mesophile Aroma B starter culture is another mesophilic starter culture, but this one is a mix of several different mesophilic lactic acid bacteria, including a few subspecies of *Lactococcus lactis* (*Lactococcus lactis* ssp. *lactis*, *Lactococcus lactis* ssp. *cremoris*, and *Lactococcus lactis* ssp. *diacetylactis*) as well as *Leuconostoc mesenteroides*. *Lactococcus lactis* is one of the primary bacteria used in the production of buttermilk and cheese. In fact, the state legislature in Wisconsin, the top cheese producing state in the country (there's a reason why Packers fans are cheese heads), tried to make it their state microbe. They would have been the first state to do so, but the senate apparently shot it down in 2010. And now Oregon has beat them to the punch. They made *Saccharomyces cerevisiae* the state microbe in 2013. *Lactococcus lactis* ssp. *diacetylactis* is a subspecies that produces acetoin and diacetyl, 2 compounds that have a buttery flavor, not surprising since they're the compounds that give butter its flavor. The *Leuconostoc mesenteroides* is a heterofermentative lactic acid bacteria that produces lots of flavor compounds and is often used in conjunction with lactococcal strains in fermented dairy production. The lactococcal strains are the main acid producers, allowing the *Leuconostoc* to do its thing and add flavor to the ferment. Trust me, it's a good combination of microbes that makes a thick, creamy, delicious crème fraîche.

Part IV

FERMENTED BEVERAGES AND ALCOHOL

Chapter 16

PAST

—

BEER

Beer production is arguably the oldest technology developed by humans. We have been making beer for thousands of years now. Besides the adoption of stainless steel and automation into the process, the techniques employed haven't changed all that much in the intervening time. If it ain't broke, I guess. Although I could argue that hazy India pale ales (IPAs) are the first signs of a fissure.

Beer and Civilization

Not only has beer been around for thousands of years, it may even be the reason why we have civilization in the first place. The Epic of Gilgamesh, from the third millennium BCE, is one of the earliest works of literature, and one that most of us have not thought about since high school. So, you probably don't recall the character Enkidu, who was the wild man created from clay by the gods to stop Gilgamesh and his oppressive ways (spoiler alert: they end up becoming friends, but that's not the point of my historical meanderings), was civilized by the prostitute Shamhat before confronting Gilgamesh in Uruk. Here is a telling passage from the epic poem:

> "Eat the food Enkidu, [it is] the luster of life. Drink the beer as is done in this land." Enkidu ate the food until he was sated; of the beer he drank seven cups. His soul became free and cheerful, his heart

rejoiced, his face glowed. He rubbed . . . ; his hairy body. He anointed himself with oil. He became human.

While seven cups, depending on the size of the cups, seems a bit excessive, especially when one is becoming civilized, it should be noted that beer from that era would have been similar to an alcoholic gruel, with an alcohol concentration no greater than a few percent by volume. So, Enkidu probably still had some of his wits about him. The important point, though, is that the story is likely an allegory for the beginnings of civilization, where wild men became civilized through the formation of organized society. Enkidu was wild until he drank the beer of civilization. Then he became truly human.

We don't have to rely on my chemist's interpretation of epic poems to inform us of the importance of beer to the founding of civilized society. There is considerable scientific and archaeological evidence to support the fact that beer, not bread, is the impetus for civilization. Importantly, we're just different from other animals. Many animals have the alcohol dehydrogenase enzyme that is the first step in breaking ethanol down to remove the toxin from the body. This is what produces acetaldehyde, another toxin that causes the flushing associated with alcohol consumption as well as hangovers (that's right, it's not the congeners in cheap liquor, although they may play a small role, or the sugar in sweet alcoholic drinks, despite common lore). Fortunately, the acetaldehyde is fairly rapidly converted as well in a metabolic process that ultimately converts the aldehyde to carbon dioxide and water.

Modern humans happen to have a supercharged alcohol dehydrogenase enzyme, which we have creatively called alcohol dehydrogenase 4; it is produced by the ADH4 gene and is even more efficient at removing alcohol from the human body. Yay! One theory is that when our ancestors came down from the trees (because they couldn't take those darn monkeys hootin' and hollerin' any longer) to begin our bipedal existence, one of their major food sources was the overripe fruit that had fallen from the trees. Well, overripe fruit will naturally begin to ferment from the yeast and bacteria found on the skin, and it will produce alcohol. Those ancestors who had mutated to have the supercharged alcohol dehydrogenase enzyme could process the alcohol faster and could thus avoid being drunken idiots all the time. They would be the more attractive mates who reproduced more successfully, thus spreading their more alcohol-tolerant genes to the next generation. That's right, young men. Despite notions to the contrary, excessive alcohol consumption does not make you more attractive. Sorry. And while this is certainly not direct evidence that beer led us to civilized society, we may have evolved to tolerate moderate alcohol consumption better than most other animals. Alcohol tolerance would certainly be helpful when trying to construct complex social structures and modern civilized societies, especially when the humans in question are consuming alcohol at the same time. Because they certainly were and have been for a long time.

The archaeological evidence for beer and other alcoholic fermentations from predominantly hunter-gatherer societies seems to mark the transition to a more agrarian society. The earliest evidence we have for beer brewing is from the Natufians, a hitherto hunter-gatherer culture, at the Raqefet Cave in modern day Haifa, Israel from about 13,000 years ago. This predates the domestication of cereals in the Fertile Crescent by several millennia. It appears that the brewing was done for ritualistic purposes, as the cave was a burial site for the Natufian people (Li Liu 2018).

The ritualistic nature of brewing back then is further supported by the fact that, despite the similarities in production of bread and beer, the technology available at the time (stone mortars and smaller stones for grinding) made the process quite labor intensive. This presents a pretty risky scenario if you're trying to use the grains for your regular dietary needs. The number of calories burned preparing the bread would probably outstrip the caloric intake from the product. That's not exactly a sustainable nutritional practice (unless you were trying to lose weight—and I'm going to go out on a limb and say that's not what the Natufians were concerned with). Not to mention the fact that, prior to widespread cultivation, which didn't happen for several thousand more years, and despite our simplistic characterization of the region as the Fertile Crescent, grain availability was spotty and dependent on seasonal growing conditions, droughts, floods, and numerous other factors that would have limited any sort of dependence on grain as a food staple. If the grains were collected on a large scale and stored, those stored grains would be susceptible to rodents, insects, and mold and would thus be too risky to depend on for regular dietary needs.

In short, the archaeological evidence points to beer brewing predating cereal grain cultivation and possibly being the impetus for widescale cultivation. The Natufian site appears to be a burial site, and the beer was likely for ritualistic purposes. However, other evidence indicates that some pre-agrarian societies traveled over a thousand kilometers to share their beers with other societies. Was this for ritualistic purposes? Quite possibly. Or could these have been the first homebrew competitions? We may never know. But given the fact that human nature probably hasn't changed much over the last several thousand or so years, it's not hard to believe. This is the way I see the development of modern civilization with my vast archaeological experience (none). These small bands of hunter-gatherers combed the region for cereal grains with which to make their proto-beer. They traveled to neighboring tribes to share their wares. Everyone sat around the campfire at night enjoying the beery gruel. The next morning, the effects of over imbibing taking their toll, they decided to stay and milk their hangovers until it was time to do it again. The next thing you know, you've got New York City. OK, so I may have skipped a couple of generations, but that's the way I see it going down with our early ancestors.

To say that beer was important to early societies would be an understatement of historic magnitude. The oldest recorded recipe is a recipe for beer, in the form of a 4,000-year-old Sumerian poem, the "Hymn to Ninkasi," the patron goddess of beer. Yup, they had a goddess of beer, which should tell you something about the importance beer held in early societies. The recipe, which was recreated in 1991 by Fritz Maytag, founder and former owner of Anchor Brewing Company, consisted of soaked grains mixed with bread and water. The bread likely harbored yeast and served as the inoculant for the proto-beer. Beer was even used as currency back in the day. The Code of Hammurabi, laid down by King Hammurabi, who reigned over Babylon from 1792 to 1750 BCE, was a set of laws that established beer rations for citizens based on their social status. Laborers received 2 liters a day, whereas priests and administrators received 5 liters, which seems a bit unfair, and only goes to show you that my point about human nature not changing is probably spot-on.

Beer History in America

Now, fast-forward a few thousand years to North America and the settlers of the American South. Even after all that time, beer remained a critical part of everyday life. In fact, beer was drunk pretty much all day, every day, by everyone, including the children. It's not that everyone turned into complete and utter lushes (although I suspect that most people went through most days with a decent buzz, which probably helped mitigate the miserable conditions that people faced in their daily lives). It's that beer was safe to drink, while water, particularly in cities, would likely kill you, or at least make you wish you were dead. Before we understood germ theory, the number of waterborne pathogens in untreated water that had been exposed to rotting corpses and fecal matter must have been downright astounding. Beer, by contrast, is safe from human pathogens due to its alcohol content, low pH, the antimicrobial properties of hops, and the fact that it is boiled before fermentation.

To be clear, it's not like people were slamming big, boozy IPAs all day. The British had come up with a brewing technique known as the parti-gyle method for brewing. With this method, a single mash (the milled grains mixed with hot water to activate specific enzymes in the grain) is used to make two and sometimes three different beers. The runoff from the first mash is sent to the boil kettle and used to make higher gravity beers like barley wines and old ales because of the high concentration of sugars collected in the first runnings. Then the mash tun is filled with water again, and the remaining sugars are extracted from the grains. The runoff from this mash would be boiled and used to make more standard beers like bitters and milds. This could then be repeated a third time to collect the last bits of sugar remaining in the grains. Because there would be so little sugar at this point, the beers made from this third batch would be small beers, totaling no more than 1–2% alcohol. These are the beers that would be consumed on a regular basis during the day and by children. Using this method, people could consume a beverage that was safe to drink and would not get them completely schnockered.

All this meant that access to fresh beer for the settlers was critical. Since people had no idea what microbes were back then, they didn't know why the water was dangerous to drink; they knew only that back in Europe, water was a no-no. Why would the pristine waterways of North America be any different? It was just nasty water, after all. Which is also why bathing and personal hygiene in general were rare occurrences at best (again, a justification for being a bit tipsy pretty much all the time). Before the settlers even departed for America, they had to ensure that adequate supplies of beer were available for the long journey across the Atlantic. Ideally, there would be enough beer for the journey, additional stores to tide them over for a bit once they made it, and adequate supplies for the captain and crew on the journey back.

This was not always the case. Famously, the Mayflower, on its journey over in 1620, was supposed to land on the shores of the Virginia colony. However, due to some navigational errors, the ship ended up much farther up the coast in New England. Once the captain and crew realized their error, why not just turn south and put ashore where they intended? Well, for one thing, they got a late start and didn't reach the American coast until December, not a time for pleasure cruising around the North Atlantic. Second, and arguably more importantly, they were almost out of beer! William

Bradford, the leader of the settler party, wrote, "We had yet some beer, butter, flesh, and other victuals left, which would quickly be all gone; and then we should have nothing to comfort us" (H. S. Rich 1974, 179; Smith 1998). Apparently, the lack of beer sealed the deal for them. "So in the morning, after we had called on God for direction, we came to this resolution—to go presently ashore again and to take a better view of two places which we thought most fitting for us; for we could not now take much time for further search or consideration, our victuals being much spent, especially our beer, and it now being the 19th of December" (H. S. Rich 1974, 179; Smith 1998). Funny that I don't remember hearing the part about the beer when we learned about the settlers landing at Plymouth Rock in grade school.

The importance of beer to the early settlers cannot be understated either. A brewery or public house was one of the first buildings constructed in any colonial settlement. Shipments from Europe of beer or the raw materials to make beer could not be consistently relied upon. So, the settlers turned to local production and used ingredients native to their new locale. The latter was particularly important in the South, where the climate was too hot for decent barley production. As Captain George Thorpe recounted of the situation in the Virginia colony in 1620, "Mr. Russell, the chemist, tried to introduce sassafras tea into Virginia as an artificial wine in July, 1620.... There is in Virginia, or is likely to be shortly, three thousand people. And the greatest want they complayne of is a good drinke—wine being too dear, and barley chargeable, which though it should there be sowen, it were hard in that country, being so hot, to make malt of it, if they had malt, to make good beer" (H. S. Rich 1974, 121; Smith 1998). If you can make it past the bizarre spelling and grammatical syntax, two things jump out here. One, Mr. Russell was apparently trying to give all chemists a bad name. And two, the South was too darn hot for decent barley production. Fortunately, the Algonquian, Iroquoian, and Siouan peoples, despite not having any fermented alcoholic beverages of their own to speak of, were generous enough to show the settlers some decent starchy alternatives, the primary one being corn. The first recorded evidence of settlers using corn in brewing dates back to 1584 along the coast of what was to become the Carolinas and Virginia (Baron 1960; Smith 1998).

The settlers at Plymouth were apparently similarly introduced to corn as an alternative grain to be used in brewing. The Wampanoag tribe, who were native to that region, fortunately took pity on the clueless settlers and taught them to hunt, fish, and gather the native, edible plants, including corn as an alternative grain to be used in brewing. In return for this kindness, the settlers famously laid out a big, post-harvest spread in 1621 for everyone; and this eventually became celebrated annually as Thanksgiving.

Other native alternatives were explored as well. This is where we get modern day pumpkin ales that nowadays represent a fun seasonal beer option, but that back in the day were evidence of the settlers using any and all available sources of starch. Another source was apparently chestnuts, which makes sense, since back before the fungus that wiped out pretty much the entire population, one in four trees in the Appalachian forests was an American Chestnut. Chestnuts apparently have a similar flavor to barley when used in brewing. The problem with pumpkins and chestnuts, however, is that they lack the enzymes to convert their starches into fermentable sugars, so an amount of malted grains would have to be added to the mash along with them (the same issue would apply to the corn as well, unless it was malted, which it apparently was in some cases). More likely, these starch sources were used to complement and stretch the limited stores of malted barley and wheat that had been imported or produced in more northern colonies.

The first permanent settlement in America was established at Jamestown, Virginia, in 1607. Unfortunately, brewers were missing from the original settlement party. Understandably, this made for a tense first couple of years before the colony could bring over two brewers to try to rectify the noticeable lack of beer in the settlement. The settlers planted barley and also imported malt and beer; but even so, the barley didn't grow so well in the South, and the imports were limited at best. So, they also turned to alternative sources of starch to stretch their brewing as much as possible.

The colony of Georgia was faced with a different problem. When James Oglethorpe established the colony in 1733, he knew that the way to a settler's heart was through beer. However, he didn't want a bunch of drunk layabouts who didn't work to establish the colony. To this end, he provided each settler with 44 gallons of good English beer. His hope was that this would help to keep them hydrated and nourished and prevent them from turning to the harder stuff. Unfortunately, he also supplied each settler with 65 gallons of molasses, which they quickly fermented and distilled into "demon rum." This led to a ban of hard liquor in the colony in 1735. Oh well, nice try, Jimmy.

As the colonies grew, so too did their desire for beer. Even the colleges got in on the game. William and Mary College in Virginia opened a brewery on its campus in 1693. When a fire tore through campus in the early eighteenth century, the brewery was one of the first buildings that was rebuilt.

Maryland, my original stomping ground, which, yes, is south of the Mason-Dixon line and therefore a southern state, is more than just crab cakes and football. No, wait, that wasn't my point. My point was that Maryland was a little late to the game, apparently. Their first major commercial brewery did not open until 1748, when a ship's captain (not surprising, considering that Baltimore was a major port even back then), Daniel Barnitz, established the John Leonard Barnitz and Elias Daniel Barnitz Brewery on the corner of Baltimore and Hanover Streets. The brewery remained in operation, under various owners, names, and even a few different locations, for over 200 years. The name was changed to the Globe Brewing Company at the end of the nineteenth century and kept that name until it was shuttered in 1963.

A Revolution Brewing

Despite the continued expansion of brewing in the colonies, the increasing volume of beer was not enough to slake the thirst of the increasing number of colonists. Fortunately, the South had something that mother England, with all her delicious beer, wanted as well—tobacco! The English had developed quite a liking (addiction) to tobacco but were unable to grow it in their gloomy climate. So, the Southern colonies in the early eighteenth century established a barter system with England that provided 1 gallon of beer for every 40 pounds of tobacco. That seems like a bad deal for the South to me, but I guess that was England's schtick. How did that turn out for them? Oh, yeah. The Northern colonies, which enjoyed a far greater number of breweries and overall production capacity, supplemented the English shipments of beer with shipments of their own to the Southern colonies.

The dependence on English imports, particularly on beer, led to increasing tensions between the motherland and her North American colonies. The ongoing war with France for supremacy over North America drained English coffers. To recoup the vast sums lost to its war efforts, England levied increasingly burdensome taxes on colonial imports from England. If they had only left beer alone, we might still call cookies biscuits and add unnecessary "u's" after every "o." That may seem an extreme statement, but beer was such a vital part of life to the colonists that most households' greatest expense at the time was beer. So, when taxes continued to increase on imported beer, it hit colonists' purse strings hard. And then things really came to a head when the Crown introduced the Townshend Acts of 1767. These acts not only levied taxes on imported goods from England, including especially beer; they also allowed for search and seizure of private businesses by representatives of the Crown. The seeds of revolution were officially sown. The Boston Tea Party, which most people attribute as at least the symbolic start to the American Revolution, was fueled by many a tankard of ale at the Green Dragon Tavern in Boston. I think only the Brits would get so upset over the dumping of a little tea down the proverbial drain. But, boy, did they.

The Boston Tea Party elicited a response from the Crown that came to be known as the Intolerable Acts. These included, among other things, the Boston Port Act, a blockade of Boston port by the British Royal Navy, which prevented any ships from entering or leaving the port, an act meant to starve the insurrectionists into submission. It had exactly the opposite effect. Prior to these intolerably harsh acts by the Crown, other colonists may have been sympathetic to the Bostonians but may also have thought the Bostonians were overreacting a bit. Now, the other colonies, including those in the South, rallied to support their friends in Boston.

The Virginian George Washington was a veteran of the British Army and a leader in the Virginia colony. Upon hearing of the Boston Port Act, Washington and a few of his best mates in the Virginia assembly, Patrick Henry, Richard Lee (no relation to Geddy—he's Canadian), George Mason, and Thomas Jefferson declared the day the act went into effect as a day of "fasting, humiliation, and prayer." For some reason, this angered the loyalist Virginia governor, who disbanded the assembly. No worries, Washington and colleagues thought, and moved their deliberations to a more favored spot at the Raleigh Tavern (where there was beer). There, they made several proclamations, including that they considered the aggression by the British on Boston to be an attack on all British America and that they would no longer import beer, ale, porter, or malt from Great Britain. A similar proclamation, the nonconsumption agreement, which would end imports from Great Britain on December 1, 1774, was made shortly thereafter at the first Continental Congress in Philadelphia. These proclamations were undoubtedly difficult for all the colonies, but the lack of imports was particularly difficult for the South. From the beginning, the South had relied most heavily on imports of beer from England. As a result, they had far fewer local breweries to fill the void. With an impending war with England, beer deliveries from the Northern colonies were few and far between. In an effort to rectify the situation, the Virginia legislature recommended an increase in hop and barley farming to increase the production of local beer.

Once the war began in earnest, one of the greatest challenges facing its leaders was the procurement of a daily ration of beer for the soldiers. The continued embargo on imported beer did not help the supply issue, but it did eventually help to establish the American brewing industry, as more local breweries opened to fill the beer void. Washington even considered allowing a ration of whiskey to replace the increasingly hard-to-find beer, as whiskey was abundant at the time. However, he thought better of providing the more intoxicating spirits to his troops and did everything in his power to keep the beer flowing. Beer prices at the sutlery were reduced for the soldiers, while at the same time, spirits prices were increased. Additionally, any outside liquor sales were banned within several miles of the soldiers' camps. Some creative solutions were devised as well. Ben Franklin shared a recipe for a spruce beer that used spruce tips to replace the hard-to-find hops at the time. Another alternative that gained popularity during this time, and which was viewed as equally healthy to beer, was cider.

CIDER

Cider production was already quite popular in Europe and had been for quite some time. The Celtic Britons were apparently fermenting cider from crabapples as early as 55 BCE (Watson 1999), when Julius Caesar and his Roman army visited the island for the Roma versus Liverpool Champions League match but were repelled by the soccer hooligans (my memory of world history is a bit fuzzy at this point, so I'm not sure that's entirely accurate—it may have been, in fact, the Roma versus Manchester United match). There is also evidence of hard cider production in the Basque country of northern Spain around this time, a natural development in a region with a strong winemaking tradition and lots of apple trees. Though I have a feeling that people had probably stumbled upon an early form of hard cider long before either of these occurrences. Wherever there were apple trees, it seems that the fermentation of the juice would be a pretty natural occurrence.

Speaking of which: apple trees were not native to North America; but when they were transplanted there, they took to their new environs quite readily. The first trees were planted in Massachusetts Bay a brief 9 years after landing at Plymouth (Orton 1995). As a result, the propagation of apple trees throughout the colonies spread rapidly. You may have heard of certain characters like John Chapman, aka Johnny Appleseed, for example. The made-for-TV version that we all grew up on had Johnny Appleseed (that loveable proto-hippy with no shoes, a pot on his head, and a burlap sack) spreading delicious, healthful apples up and down the Ohio River Valley. Apparently, this representation of his notorious fashion sense is somewhat accurate. Otherwise, though, the story leaves a bit to be desired. Johnny was religiously opposed to grafting apple trees onto existing stock. I have no idea what the religious objection to grafting is, nor do I think I want to know; but what this means is that Johnny was using apple seeds to spread his apple nurseries along the frontier (hence the whole nickname thing).

The problem with this propagation method is that apples are extremely heterozygous, meaning that the seeds contain a lot of genetic diversity. Any one seed that grows into a tree will not be of the same variety as any other seed from the same tree or even of the parent tree. This confers an evolutionary advantage on apples. If they spread through their seeds, the greater diversity of the many genetic variants that grow means that there is a greater likelihood that at least one will be suited to its new environment and survive and thrive. For apple growers, by contrast, this means that grafting is the only way to guarantee which variety you end up with. So, the nurseries started by Johnny's apple seeds were filled with completely different varieties of apple trees, the vast majority of which looked nothing like the apples we see in the grocery stores these days. In fact, very few apples back in the day looked anything like the ones we have today. You see, people didn't traditionally eat apples. But they did ferment the juice and drink the results. Most apples would have resembled what we refer to nowadays as crab apples, which are the same genus as "regular" apples, but with fruit that is less than 2 inches in diameter. And they probably were almost universally more bitter or acidic than sweet. Which wouldn't have made for great eating, but which was perfectly fine for hard cider production. So, our proto-hippy, religious missionary friend Johnny Appleseed wasn't spreading healthful apples to the frontier to keep the doctor away; he was spreading the raw materials for making alcohol.

In fact, the whole apple-eating thing didn't start for several more decades when, during Prohibition, the apple industry had to think of new ways of getting people to buy their product. If they couldn't make hard cider, whatever could they do with them? Some industrious apple growers came up with new hybrid varieties that were bigger, plumper, juicier, and especially sweeter, which made them more tempting to, you know, eat. And they came up with new marketing slogans, like "An apple a day keeps the doctor away," and "as American as apple pie." Before our obsession with perfectly plump fruit with universally acceptable flavor (i.e., boring, yeah, I'm looking at you Red Delicious and Granny Smith), the spread of cider apples to the frontier was a welcome addition to the raw ingredients available for alcohol production. With the constant struggle of sourcing beer ingredients, cider became extremely popular in the colonies.

The colonists also discovered a nifty little trick to make their cider-drinking experience even more intoxicating—jacking. No, they weren't getting drunk and horsejacking (the colonial equivalent of carjacking). Jacking is a colloquialism that refers to ice distillation. When barrels of cider were left out in the freezing cold, the water would freeze before the ethanol. If the ice was removed from the barrel, what was left was a concentrated form of cider with an alcohol content of roughly 25–40%. This ice-distilled spirit was referred to as applejack, a form of apple brandy created through the nontraditional means of fractional crystallization, or freeze distillation.

As it was first produced by William Laird in colonial New Jersey in 1698, it also became known as Jersey Lightning. However, in the South, it was and is still referred to as applejack. Laird's great-grandson, Robert Laird, founded Laird's Distillery in 1780 in New Jersey, and Laird and Company in Scobeyville, New Jersey, is still in operation today. In fact, until a couple decades ago, it remained the sole remaining producer of applejack in the country. At present there still aren't many, with most of them concentrated in the Northeast or Midwest, where some of the biggest apple producers are located; but there are several besides Laird and Company. There is even a producer in North Carolina. Holman Distillery—located in Moravian Falls, which is just southeast of Wilkesboro, the former moonshine capital of the country—makes a few different products, including a high-proof vodka; Holman's Applejack, which is actually a distilled apple brandy; and Applejohn, which is actually applejack, and supposedly the only true freeze-distilled applejack remaining in the country. Applejohn is just under 20% alcohol, so about two to three times the strength of hard apple cider but only half the strength of typical apple brandy. One concern when jacking is that there is no control over what gets concentrated when the water is removed. Sure, the ethanol is concentrated, but so too are other compounds, the most worrying of which is methanol. When the product is still lower in alcohol concentration, like Holman's Applejohn, this isn't as much of a concern. However, for a more concentrated product, a molecular sieve to remove the methanol should be used to ensure a safe product. As Holman's has demonstrated, most of the "applejack" products available nowadays are actually apple brandy produced through the typical evaporative distillation method.

Let's hope that we see more applejack and apple brandy distilleries open up in the South in the future. North Carolina (number 7) and Virginia (number 6) are the two southern states in the top 10 for apple production. Henderson County in North Carolina, just south of Asheville, is responsible for 85% of the state's apple production. Every fall around harvest season, they host the North Carolina Apple Festival, which is a fun time for the whole family if you can deal with the traffic in and around Hendersonville during the festival. Interestingly, though, apples are not the state fruit of North Carolina. That distinction belongs to the scuppernong grape, which demonstrates the historical and current importance of the grape and its most valuable product, wine, to this region.

WINE

When you think of the American South, wine may not be one of the first things you think about (seems to be a recurring theme, huh). Well, it probably should be. Kentucky and Virginia are both among the top 10 states with the greatest wine production, with Texas and North Carolina just out of the top 10, and Florida and Tennessee not too far behind. The South is also home to the state with the fewest wineries, which would be Mississippi, with a whole 2 wineries as of 2023.

North Carolina started it all and used to be the top wine-producing state. Even as recently as the early twentieth century, North Carolina was the number one wine producing state in the country. And then came Prohibition. Most of the wineries shuttered during that time, and the vast majority of alcohol production was shifted to stills hidden in the woods in the High Country. Only in the last couple of decades has a renaissance of wine production in the South occurred. Long before then, though, it all started with one vine. Not really, but the Mothervine, as she is called, located on the northern end of Roanoke Island near Manteo, is a massive scuppernong vine, one of the varieties of muscadine grape, *Vitis rotundifolia*, that is still thriving and may have been doing so since 1584. When Philip Armadas and Arthur Barlowe came to the island during their American expedition, funded by none other than Sir Walter Raleigh, they remarked on the abundance of grape vines growing there, likely cultivated by the Croatans who lived there at the time. Was one of them the Mothervine? Possibly, but this is not confirmed. We do know that she has been around since at least the 1720s, though. Peter Baum received a land grant during that time that included the area where the vine grows. A relative of Peter Baum who grew up in the area in the early nineteenth century recalls his grandfather talking about the size of the vine when he was a child. So, it was likely already established when Peter Baum received the land grant and had probably been growing for quite some time already.

It's important to note that the native grapes growing in North Carolina were *Vitis rotundifolia*, or muscadine, which are still quite popular in our neck of the woods. In fact, Duplin Winery—with its primary location in Eastern North Carolina and additional locations in South Carolina and Florida—that produces exclusively muscadine wines, ranks in the top 50 for largest wineries in the country. I'm guessing that most readers of this book who are from the South have tried some variety of muscadine grape or wine. There are certainly plenty to choose from, with more than 150 muscadine cultivars grown around the region. They all generally produce smallish berries with thick skin and have a similar and unique flavor profile, which some say is reminiscent of burned rubber. We'll just stick with "unique." It's definitely an acquired taste, or something that you need to grow up with to truly enjoy. I would say that they're not my favorite, but as a citizen of the South, I am legally forbidden from professing anything but love for the muscadine. However, let's just say that there's a reason why most muscadine wines are on the very sweet end of the wine sweetness chart.

Now that I have unsuccessfully navigated those waters and incensed half the population of the South, let's talk about a more uplifting subject: how the native American grapes saved the global wine industry. The vast majority of what we consider to be fine wines, or even most not-so-fine-wines, come from *Vitis vinifera* grapes. These are your classic Cabernet Sauvignon, Pinot noir, Riesling, and Chardonnay grapes, along with many, many more. There are many grape vines native to North America, but none of them is *Vitis vinifera*.

Two of the predominant native species, at least in the southeastern United States, are *Vitis rotundifolia*, which actually has its own subgenus, *Muscadinia*, and *Vitis riparia*, the riverbank (hence the Latin, *ripariu*) or frost grape. *Vitis vinifera* grapes are likely native to the South Caucasus region—which encompasses modern-day Armenia, Georgia (the country, not the Bulldog state), Azerbaijan, northern Iraq, and northwestern Turkey—and were spread west across Europe as early human civilization spread to that area. They undoubtedly developed tolerances for plenty of pests and diseases in that part of the world as they evolved together over the ages. However, they were not ready for one particular North American pest that made its way across the Atlantic in the late nineteenth century, a result of increased globalization at the time. Phylloxera are a small, sap-sucking root louse that is related to the aphid; it attacks both the root system of a grape vine as well as the leaves, creating galls under the leaf, ultimately causing enough damage to kill the vine. Since the grapes native to North America evolved along with these little vinous vampires, they developed defenses against the puny predators. Namely, the native grape vines excrete a sticky sap through their roots that chokes the insect nymphs when they try to feed on them; and the native vines are also able to quickly grow a protective layer of tissue on any wounds that the insects create, thereby preventing any further damage to the area. *Vitis vinifera* vines, by contrast, developed no such defenses, since they did not evolve with the minuscule munchers. So, when they were inadvertently exposed to the tiny terrors in the late nineteenth century, the European wine industry was almost wiped out.

Phylloxera were first discovered in 1863 in a greenhouse in Hammersmith, London, and were ultimately named *Phylloxera vastatrix* in France in 1868. They have since been reclassified as *Daktulosphaira vitifoliae*, but that's quite the mouthful. They are generally just called phylloxera. The French got involved then because the pint-sized pests were found in vineyards in the Rhône valley of France. They were then discovered in Bordeaux by 1869, and not even a decade later, by 1877, they had already arrived in Geelong in Victoria, Australia. Soon after, in 1885, the herbivorous horrors were found in New Zealand. They may not have spread as quickly as they would today with air travel, but considering the expansion of the wine industry around the world at the time, they had no problem tagging along for a global expansion of their domain.

It cannot be stated vehemently enough that this era was a terrifying time for the wine industry. *Vitis vinifera* grapes, which were, and still are, responsible for the vast majority of wine production around the world, were almost wiped out by the ferocious phylloxera. Tragedy often leads to innovation, however, which was exactly the case in this situation. It was discovered that grafting *Vitis vinifera* vines on top of the American species *Vitis riparia* would protect the roots of the vines against phylloxera attack and allow the aerial portion of the vine to grow and thrive. This was the innovation that saved the wine industry. The American species, with its defenses against the diminutive devils, provided a viable root stock so that the *vinifera* species, grafted on top, could still grow the grapes desired by winemakers to produce traditional varieties. Other American vines were also used successfully as rootstock, including *Vitis berlandieri* and *Vitis rupestris*, among others.

Despite this successful innovation that saved wine as we know it, the effects are still felt to this day, more than a century and a half later. Most vineyards around the world have to purchase grafted vines from nurseries, which is considerably more expensive than using ungrafted vines. Even in the few remaining areas around the world that are still phylloxera free, which include areas in Chile, Australia, Germany, and Argentina, great care must be taken through rigorous quarantine practices, such as sterilizing farm equipment and new planting material (pretty much anything that will come in contact with the vines). Additionally, phylloxera are not too fond of sandy soils, so wine regions with particularly sandy soils provide greater protection against the minute menaces.

As I mentioned above, it would take more than a petite plague to take down the wine industry in our country, however. That's right, it was no disease or pest outbreak, it was something far more insidious, the scourge that we call politics. North Carolina's thriving wine industry in the late nineteenth and early twentieth centuries was effectively wiped out by Prohibition. The irony is that Prohibition started out as the temperance movement, which was pretty much what it sounded like. For the most part, people didn't have problems with beer, wine, and cider. Those were considered healthful beverages that didn't lead to the domestic and health-related issues attributed to distilled spirits. The temperance movement really only wanted to cut down on the overconsumption of the hard stuff.

Unfortunately, as the world devolved into madness leading up to World War I, those healthful beverages, particularly beer, became associated with the German aggressors and, once coupled with the matter of resources being diverted to wartime efforts, the baby ended up thrown out with the bathwater. Temperance shifted to outright prohibition, and all alcohol was verboten. Despite that, people still drank. And the illicit producers weren't making fine wines or well-crafted beers. If producers were going to put their lives and livelihoods on the line, it was going to be for a product that gave them the biggest bang for the buck—distilled spirits. Likewise, consumers were looking for the cheapest form of concentrated alcohol they could find. So, one industry's downfall was another industry's boon I suppose. The downfall of legitimate alcohol industries in the country led to the explosion of the illicit moonshine industry, centered smack-dab in the South.

MOONSHINE

North Carolina was arguably the center of the infamous Moonshine Belt comprising Alabama, Georgia, Mississippi, North Carolina, South Carolina, Tennessee, and Virginia. Franklin County, Virginia, and Wilkes County, North Carolina, were locked in a historic battle for supremacy as the Moonshine Capital of the United States. Apparently, though, everybody knew it was really Wilkes County in North Carolina (Stephenson 2017), which is not surprising if you're familiar with the area. I'm not sure why I feel a strange sense of pride about this either. I guess I've been living in North Carolina for long enough to feel pride in even the most dubious of honors for the state.

Distilled spirits have been produced in the South ever since European settlers moved here. The settlers brought their traditional knowledge and techniques but had to adapt them to their new environment. One of the main adaptations was the use of corn, the primary cereal grain crop of the Americas. Cereal grain–based distillations in Europe involved primarily barley and wheat and possibly some other grains, but not corn. For farmers back then, distilled spirits represented a serious value addition for an otherwise basic commodity like corn. Furthermore, raw corn takes up a lot of space. So, they could make considerably more money from a product that could be transported far more easily and economically than the raw material. It was pretty much a no-brainer for farmers back in the day. Not to mention that life was physically hard back then—something that could be assuaged with a couple of nips of rattlesnake milk.

Before the Civil War, there were no laws preventing the production of distilled spirits, and the spirits were not taxed (except for the "whiskey tax" implemented by Washington in 1791 to pay the debts from the Revolutionary War—the tax was later repealed by Thomas Jefferson in 1802). But then the Civil War happened, and the government suddenly needed money to fund a war. Then came the taxes. On July 1, 1862, Congress passed the Revenue Act, which created the office of Commissioner of Internal Revenue and enacted what was supposed to be a temporary income tax to pay for the war. Well, we all pay income taxes to this day, so you can fill in the dots for the rest of that story. As far as the swamp gravy was concerned, the first taxes were imposed in 1863 at $0.20 per gallon. And suddenly, nobody was allowed to make distilled spirits without paying the tax. OK fine. But it didn't stop there. Taxes continued to increase until the government realized that they

Moonshine still inherited from the shuttered
Appalachian Heritage Museum.

were probably actually missing out on a good bit of tax revenue because of all the distillers who had gone "underground" because they couldn't afford the tax burden. So, in 1868 the tax was reduced again, and distillers were provided a grace period of one year on their tax obligations, referred to as the "bonding" period. All product had to be stored in government regulated "bonded" warehouses during this time. This was due to the fact that prior to 1868, distillers had to pay taxes on the spirits once they ran out of the still, even though many would age their products for years. So, for those that did age their products, they were forced to pay taxes long before they were able to recoup their money. Well, one year is nothing when it comes to aging spirits, and the government was under pressure to support an industry that provided a large percentage of the overall tax revenue it collected. As a result, the bonding period was increased incrementally until the Forand Act of 1958 increased it to 20 years (Regan and Regan 2009).

Still, the damage of excessive taxation over the previous century or so had driven many distillers to abandon their legal enterprises and continue them instead in an illicit fashion. Many distillers moved their stills to the dense forests around the South, away from the prying eyes of government agents. The term "moonshine" likely derives from the fact that distillers, when they were distilling in the deep woods, often did so at night to further evade detection; they had to work by the light of the moon, or the moonshine.

Taxes on distilled spirits are one thing. But what about the complete prohibition of any alcoholic beverages? Yup, we tried that too. That "Noble Experiment," national Prohibition, lasted from 1920 to 1933; it resulted from the ratification of the Eighteenth Amendment and passage of the Volstead Act in 1919. We all know how that went. Sure, there is evidence that alcohol-related illnesses decreased during Prohibition, but there is also ample evidence that alcohol-related crimes increased by even larger margins. Regardless (and thankfully), support for Prohibition decreased with every passing year, and it was finally ended with the ratification of the Twenty-First Amendment. But the damage to the distillation industry had already been done. Apparently, some of the largest moonshining stills ever captured by government agents were found during the Prohibition era.

Western North Carolina, the "High Country," is home to the tallest mountains on the East Coast. If you drive south along Highway 321 from Boone through Blowing Rock and gaze out to your right, you are provided with some of the most amazing highway views in the country. On clear days, you can see for hundreds of miles. What is most striking is the fact that the view of rolling mountains and dense forests is virtually unbroken by any type of development. And that is now. I can't imagine what the region must have looked like 100 years ago. The point is that this is the perfect area for moonshining—dense forest cover, plenty of fresh, clean water, and located in the middle of nowhere, with questionable (if any) vehicle access. For these reasons, the High Country, following its settlement by Scotch Irish immigrants with their knowledge of and experience in distilling, was the birthplace of moonshine. And Wilkes County, in the foothills on the eastern edge of the region, was the capital.

Wilkes County was also home to Junior Johnson, legendary Hall of Fame NASCAR driver, who cut his race car–driving teeth running moonshine all over the High Country. His father was a lifelong bootlegger who taught his son to drive long before he was eligible for a driver's license; and Junior was apparently evading federal agents in high-speed car chases long before he was old enough to drive legally. However, Junior turned to more legal uses of his driving skills in the form of NASCAR. He was not the first to do this but was arguably the most successful and popular of the bootleggers-turned-NASCAR drivers.

In fact, NASCAR got its start from bootleggers during Prohibition. Moonshiners would make their product up in the mountains, and the drivers would distribute it throughout the region. The drivers would use vehicles that were small and fast, and they would often modify them to increase their speed and handling so that they could outrun the police. Even after Prohibition was repealed, moonshiners continued to ply their illicit trade to avoid the tax obligations of legal distilled spirits. Continued production of illicit moonshine meant a continued need for drivers runnin' shine across the region. Those drivers began to race among themselves for pride and even prize money. Soon the races between the vehicles became a major source of entertainment in the South. To legitimize the sport, William France Sr., a driver and promoter of the sport himself, together with other prominent drivers and promoters, formed the National Association for Stock Car Auto Racing (NASCAR) on February 21, 1948.

WHISKEY

Uncle Nearest and the History of Tennessee Whiskey

Nathan "Nearest" Green, known endearingly as "Uncle Nearest," was born into slavery in Tennessee around 1820. He was owned by the firm Landis and Green that hired him out to Dan Call, a Lutheran preacher who operated a farm and also a whiskey still on his property, the Dan Call Farm, outside of Lynchburg, Tennessee. Nearest worked with Call in his distillery and apparently became a very skilled distiller. Around the mid-nineteenth century, Call married a teetotaler named Mary Jane (who maybe preferred other intoxicants based on her name?) who was less than thrilled with his distilling. This sentiment was apparently shared by Call's congregants at the time. As such, Call stepped back from the distilling operations and promoted Nearest to head stiller (now referred to as master distiller). He was the first African American to hold that position in the country. He may, in fact, be responsible for perfecting the Lincoln County Process, in which the distillate is filtered through sugar maple charcoal prior to aging it in barrels. This process gives Tennessee whiskey its characteristic smoothness and differentiates it from any other whiskey in the world. Regardless, Nearest made some darn good whiskey that had a reputation far and wide.

Shortly after Nearest took full responsibility for the Call Farm distillery, a young boy name Jasper "Jack" Daniel, who was attending the nearby Mulberry Training Academy, came to work on the farm as a chore boy. He befriended Nearest and eventually became his pupil in the distillery. For many years, it was incorrectly reported that Dan Call taught Jack to distill; however, a story in the *New York Times* in 2016 finally set the record straight and gave proper credit to Nearest Green, the head stiller at the Call Farm distillery. This is where the history gets a little murky. After ratification of the Thirteenth Amendment, Nearest Green, newly freed, stayed on as head stiller at the distillery. At some point, Jack Daniel took over the distillery on the Call Farm and retained Nearest as the head stiller and also employed two of his sons, George, his second-born, and Eli, his fourth-born. Then, sometime after 1881, Jack moved his distillery to Cave Spring Hollow, where it remains today. Jack brought with him a few of Nearest's children and at least three of his grandchildren as well. To this day, there have been seven generations of Greens who have worked at the Jack Daniel distillery, and there are three who work there to this day (Green 2017).

Bourbon

Captain George Thorpe—the previously mentioned early Virginia colonist who lamented the lack of good barley and the difficulty growing it in the hot, southern climate—was one of the first colonists to embrace the Native Americans' alternative grain, corn. In fact, he may have been the godfather of American distillation, the first to use corn to replace some of

the more traditional barley and rye from European distilled spirits. As an interesting aside, Thorpe was probably also involved in what was arguably the real first Thanksgiving. Thorpe's colony, the Berkeley Plantation, located just outside the main Jamestown colony, gave thanks for surviving the harrowing Atlantic crossing aboard their ship, the *Margaret*. Captain John Woodlief (was everyone a captain back then?), the leader of the colony, found a nice, grassy clearing where they spread out the remains of the ship's hold, including the aqua vitae (more on this soon), to share with the Native population. This happened on December 4, 1619, a full 1 year and 17 days before the copycat Thanksgiving in Plymouth. Even President Kennedy acknowledged this fact, but everyone else has pretty much ignored this little historical update.

Captain Thorpe went to great lengths to understand and learn from the Native population, including their agricultural practices. He embraced corn, or maize, which translates as "that which sustains us," like few other colonists before him. He incorporated it into beer, replacing the traditional cereal grains; but he also may have been the first colonist not only to use corn to make distilled spirits but also to distill in the Americas in general. There is a monument to where his purported still was located at the Berkeley Plantation outside Jamestown, Virginia. Well, it's a water fountain that looks like a whiskey barrel with a plaque, stating "First Whiskey Distillery 1621," but that counts. Right? Regardless, these first attempts at distilling a corn mash were still a long way from modern bourbon.

Rum was arguably the most common distilled spirit found in the colonies in the intervening years. Not only was the British Crown producing rum from the byproducts of their sugarcane plantations in the Caribbean, they were also shipping the raw materials, sugarcane and molasses, to the industrialized cities in the colonies. There, the raw materials were converted to the more valuable commodity, rum. However, as war with England became imminent and British imports dried up, coupled with a disdain for any products associated with England, distillers turned instead to native grains and to whiskey production.

George Washington (who seems to be a recurring character in this part on alcoholic beverages), that renowned lover of porter, was apparently quite fond of the harder stuff as well. Following the Revolutionary War and two terms as president, Washington was understandably ready to kick up his size 13 buckle shoes and retire to the good life at Mount Vernon. There, he hired Scottish emigrant James Anderson to manage the farm. Anderson convinced Washington to build a still on his property, and for a brief period

How Whiskey Got Its Name

So, about that aqua vitae. The term, which translates as living water, actually referred to distilled spirits of an unspecified nature. The term was coined by Arnaud de Ville-Neuve, a French alchemist and physician at the University of Paris in the fourteenth century. The fact that someone could be an alchemist and physician tells you all you need to know about the state of science and medicine back then. Oh, well. Arnaud, with his vast knowledge of scientific principles, was convinced that distilled spirits were the concentrated essence of sunlight, which was previously captured by the fruits and grains that were fermented and distilled. But wait, there's more. According to Arnaud, this "essence of sunlight" was "a water of immortality [that] prolongs life, clears away ill humors, revives the heart, and maintains youth." You know, I've tried to tell my wife this for years, but she's not buying it. Whereas his understanding of science was a bit limited (oh, if only he could have been correct), the name aqua vitae for distilled alcoholic spirits stuck. You may have heard of aquavit (or akvavit), the Scandinavian distilled spirit (serve with *hákarl*, fermented Greenland shark, for the true experience). The same term translates to *l'eau de vie* in French and *usquebaugh* in Gaelic. Over time, the Gaelic became *uisge-betha*, and then was shortened further to *uisge*. Eventually, *uisge* became the more Anglified whiskey, which referred to distilled spirits from fermented grains produced in the British Isles (Mitenbuler 2015).

Now that we know the origin of the name, what's the correct spelling? Is it whiskey (yes) or whisky. That depends where you are in the world. In the United States and Ireland, we spell it correctly, as whiskey. In Scotland, Canada, and Japan, they incorrectly spell it whisky. Just kidding, either spelling is correct, depending on where you are. The Japanese, with their distillation industry based on the Scottish model, understandably use the same spelling. Canada, by contrast, has no excuse. But they can't even decide what language they speak, so we'll let them slide. There is an unsubstantiated story that Scottish printers were cheap and dropped the "e" in whiskey to save money. I have no doubt the story was started in England, and likely has no basis in reality. In short, the reason for the different spellings is unknown and probably just the result of terrible spelling in general back in the day (although, as an educator, I can attest that it hasn't improved much in the intervening years). Regardless of the spelling, the whiskeys (whiskies if you're Canadian, Scottish, or Japanese) from each of these countries are unique and excellent. Now, time for a glass of bourbon whiskey! •

it was the largest still in the country. Sadly, Washington died shortly thereafter, followed not long after by Martha, and Anderson left to follow other pursuits. The still burned down in 1799, and records from its brief existence are spotty at best; but in 2007 the still was resurrected to much fanfare and foofaraw. The first bottle of the whiskey produced from the still sold for a cool $100,000 at auction to Marvin Shanken, publisher of *Wine Spectator* and *Cigar Aficionado* and all-around lover of the good life.

The Scotch Irish immigrants were instrumental in developing the American whiskey industry. Escaping oppressive whiskey taxes in their homeland, they sought a new home and independence from an overreaching government. They found the freedom they craved, particularly as they spread west and began to settle the frontier lands. These were the Ulstermen, poor Protestants who were relocated by King James I in the early seventeenth century to Ulster, a rebellious Catholic region in Ireland, to dilute the Catholic influence and spread good old English Protestantism. To make ends meet, the Ulstermen turned to distillation. This is when many of the distillation techniques were developed and perfected that ultimately led to the Scotch and Irish whiskeys we have today. They were so successful, in fact, that the Crown was none too happy. So, it increased rents and limited trade with the rest of the country. The nail in the coffin, though, was when the Crown imposed whiskey taxes to fund a civil war in another part of the country. When the option arose to emigrate and exercise their distillation expertise thousands of miles away from the oppressive British monarchy, the Ulstermen jumped at the opportunity.

The distillation practices during the seventeenth and eighteenth centuries were quite primitive but, at the same time, not too dissimilar to modern practices. The introduction of corn as one of the cereal grains used, and quite often as a primary grain, is one of the practices that began during that time and one that makes modern American whiskeys so unique and delicious. The corn was ground to a coarse meal and blended with other milled grains, particularly barley, rye, and wheat. Hot water was added to the milled grains to make the mash. Today, we understand that we do this to activate the enzymes in the malted grains (which is why a small amount of malted barley would likely have been added, since unmalted corn has no active enzymes) that will degrade the starches to simple sugars that can be fermented by the yeast. Aside from our current understanding of enzymology, however, a modern mash would look very similar to a mash of yore.

Back in the day, the mash would sit and cool and then would either be unwittingly inoculated with yeast from a prior batch using some type of transfer mechanism for the *godisgoode* (yeast), or it would be inoculated with wild yeasts from the environment. Nowadays, thankfully, the cooled mashes are intentionally inoculated with specific amounts of particular yeast strains that can best get the job done. I write thankfully because, before we had any knowledge of microbes and fermentation, animal carcasses would sometimes be added to the mash to kickstart a fermentation. Oh boy.

Once the mash was done fermenting, the mixture was distilled much like it is today. Sure, there were no column stills back then, but many modern distilleries still use pot stills. It just depends on the product being made and desired flavor profile. Most stills today are not heated with wood fires, but there are still some out there. Otherwise, the actual distillation portion of the process in the seventeenth and eighteenth centuries was not terribly different from what it is today. I would hope that distillers have a better understanding of the proper cuts to make of the heads, hearts, and tails (the fractions that come out of the still—see chapter 17 for a detailed description), but there's no reason that a distiller from back then wouldn't have mastered this technique as well.

Probably the biggest difference in whiskey production between the two eras is in the aging of the spirits. Or the lack thereof in historical practices. That's right. The raw distilled spirit would have been added to ceramic jugs or wooden barrels that were not charred or toasted but intended strictly for storage and transport, not for flavor and color addition. If stored in wooden barrels, even without charring or toasting, some flavor and a little color would undoubtedly have been picked up, but this was incidental and not an intentional part of the process. Now that whiskey is enjoyed as a luxury product and not as something necessary for survival on an unforgiving frontier, we take full advantage of the flavor and color enhancement that comes from aging the product in charred and/or toasted oak barrels. This process, which can last for years, even decades sometimes, is absolutely critical in the production of fine whiskeys.

Back in the eighteenth and nineteenth centuries, however, the burgeoning whiskey industry in the United States hinged on which political vision would win the day for the future of the country. One vision was that of Alexander Hamilton and his musical theater vision for the country. Just kidding, that bit would come much later and with very little input from the man himself. But Hamilton's vision was the more traditional one, with institutionalized taxes favoring the larger producers to squeeze out the smaller players from the industry. This part was intentional. Hamilton believed that the entrepreneurs in the frontier were much harder to oversee and lacked the necessary efficiencies of the larger, coastal businesses of a modern, industrialized society. Jefferson, by contrast, despite being a wealthy elitist himself, supported a vision for a country made up of rural, working-class farmers. He knew that Hamilton's tax scheme would not be viable for his vision of the country. When he became president and repealed the country's first whiskey tax, it was a victory for the small, independent distiller.

However, that only lasted for a few decades. Hamilton's vision ultimately won out, and the modern American whiskey industry is a direct reflection of his vision for the country and its industries. In the year 2000, there were only 13 distilleries in the country (and possibly a few tiny independent distilleries) owned by eight companies. Despite the fact that you could go to the liquor store and peruse the hundreds of whiskey brands for sale at that time, pretty much all those different brands were produced at those 13 distilleries. Craft distillation has since exploded, however, and by 2005, there were 57 craft distilleries. There are now well over 2,000 distilleries operating in the country. That said, the 13 aforementioned big plants are still probably responsible for over 90% of the total volume of American whiskey production.

Jefferson and Hamilton have both long since passed, but their symbolic battle rages on. It may seem that Jefferson's philosophy is once again gaining prominence, and to a degree, you would be correct in this observation. There are thousands more independent craft distilleries today than there were 20 years ago. Yay, Jefferson! However, the fact that 13 massive distilleries still control the vast majority of whiskey production in the country is a win for Hamilton. And, honestly, their contradictory philosophies are both quite evident in the modern industry. The modern, industrialized vision of Hamilton has certainly been realized by the whiskey industry. And the result is highly regulated, safe, and delicious whiskey for all. I'll take that. The story the industry tells the consumer is completely different, though, and one much more in line with Jefferson's vision. It tells a tale of the small, independent, and rebellious American distiller bucking the industrialized trend and making a traditional, old-timey recipe found in someone's dust covered attic that supposedly came from one of the founding fathers or a buck skin wearing frontiersman. Which is all well and good for marketing purposes. Unfortunately, most of it is completely fabricated.

First of all, Jefferson didn't even like whiskey. He was a wine guy who thought whiskey was way too unrefined for his highfalutin palate. But, at the same time, he absolutely did support the small, rural producers of whiskey. Because of that, he is still revered within the whiskey industry, while Hamilton is viewed as a Wall Street pariah. Jefferson even has his own brand. Well, he doesn't own it, obviously, but the brand is named for him. The irony is that the brand is actually contract-distilled at one of the big 13 distilleries, that is, in Hamilton's domain. And this is not a unique situation. Far from it, in fact. Many of the brands we know and love, that without microscopic investigation of the label and website appear for all the world to be independent distilleries, are in fact just brands produced at one of those same 13 distilleries. It doesn't stop there, either. Even the history of bourbon itself is veiled in the same kind of marketing sleight of hand as the brands themselves.

Elijah Craig is credited with being the godfather of bourbon. You probably also recognize the brand, which is quite good in my opinion. I just picked up a bottle of the barrel-strength Elijah Craig, and I highly recommend it. But Elijah Craig is not a distillery. It's a brand that has been distilled since 1986 at Heaven Hill Distillery, which was established in 1935. Now back to the man the brand is named for. Elijah Craig was a firebrand preacher from Virginia who was even jailed for a time for his fiery rhetoric. He later moved with several hundred of his devoted flock to the frontier land of Kentucky where he could continue his preaching with greater independence.

Craig, like many of the other early settlers, took advantage of the deep, loamy soil of Kentucky to grow an abundance of corn. Kentucky was so fertile back then, that the yield for corn was supposedly 40 bushels per acre compared to 10 bushels per acre in Maryland. What to do with all that corn? Well, make whiskey, of course. And that's exactly what Craig did, much to the chagrin of his preacher colleagues. And this is where the truthful extent of his contributions to the nascent bourbon industry ends. That hasn't stopped Heaven Hill from borrowing a bit of invented history to go with their popular brand. According to their accounts, a barn fire charred the insides of the barrels (a very selective fire, or were the outsides charred as well?) that Craig intended to use to store and transport his distilled spirits. Having no other options, Craig used the barrels anyway, and lo and behold, bourbon was born!

Cool story. If only it were true. And no fault to Heaven Hill. They're only using what has been the accepted lore in Kentucky for quite some time. In 1874, historians Lewis and Richard Collins credited Craig with making the world's first bourbon. They didn't mention Craig by name, mind you, but wrote that "the First Bourbon Whisky was made in 1789, at Georgetown, at the fulling mill at the Royal spring" (Collins and Collins 1874, 516; Mitenbuler 2015). This would identify Craig as the distiller in question.

However, there has been no evidence provided that differentiates the product Craig was making from the other whiskeys produced in Kentucky at the time. Modern historians theorize that the Collinses were attempting to paint a more respectable picture of whiskey by tying its emergence with a religious leader at the time, in an era when it was under attack by the temperance movement. Regardless, it's a good story with an interesting character. Catchier than the more sedate but accurate history that includes the contributions of numerous farmers, Indigenous peoples, and enslaved people living on the frontier at the time. That would be a much more difficult, and arguably less marketable, story to impart in the limited space on the back of a bottle.

Why has Kentucky emerged as such a powerhouse of whiskey production? Well, historical happenstance certainly played a role, as did the abundance of corn; but Kentucky's climate in conjunction with barrel aging is another reason for the emergence of Kentucky bourbon as one of the world's great whiskeys. The distilled spirits are placed in toasted and/or charred barrels and allowed to age for years in a rickhouse, a non-climate-controlled warehouse that houses all the whiskey-filled barrels.

Historically, the rickhouses could not have been climate controlled, but they certainly could be today. So, why aren't they (to be honest, some are, but only to create a controlled environment meant to mimic the seasonal changes that would normally be experienced anyway)? In fact, subjecting the barrels to large temperature swings is one of the keys to "aging" whiskey. In the summer, when it's nice and hot in Kentucky, the barrels swell, and the wood pores open up. This increases the surface area inside the barrel and accelerates the flavor enhancements to the distilled spirit as it fills the opened pores. In the winter, when it cools down, the barrel contracts and the pores close up, decreasing the surface area of the wood and pushing the spirit back into solution. Every day, every season, as temperatures change, this phenomenon occurs, creating a breathing-and-mixing type cycle of the wood and whiskey inside.

There can even be rather large temperature differences within the same rickhouse. Depending on the height of the structure and how high the barrels are stacked, the top barrels may be several degrees warmer than the lower ones. Many distilleries will rotate the barrels to take advantage of these temperature differences. Otherwise, the final products are blended to balance the more strongly flavored upper barrels and less strongly flavored lower barrels. The more frequent and drastic the temperature changes, the more flavor and color is picked up by the whiskey, thereby accelerating the aging process. This is why Kentucky bourbon can be incredibly flavorful and ready to bottle after only a few years, whereas Scotch typically takes several years more of aging—because Scotland is generally cool year-round, with much less drastic temperature swings.

The humidity plays a role as well. The more humid it is, the less water will evaporate from the barrel, which means more alcohol will evaporate, relatively speaking. When it's less humid, more water will evaporate relative to the alcohol. Since it's generally humid in Kentucky, alcohol loss can be quite large over time, which is fine for barrel aging. A lower alcohol concentration encourages greater flavor addition from the barrel, as most of the compounds are more water soluble. The evaporative loss of whiskey in a barrel is referred to as the angels' share. Which is also where Angel's Envy gets its name, because what's left in the barrel after aging is the envy of the angels.

Bourbons that age for several years will often have very little whiskey left because of all the evaporation. That's why longer-aged whiskeys can be so expensive. The porous nature of barrels also allows for plenty of oxygen ingress, particularly during cool contractions of the wood, which creates an oxidative environment that further enhances flavor development. Chemical esterification also occurs over the long aging period, in which hot fusel alcohols (alcohols with more than two carbons, the number that ethanol has, which are produced as byproducts of fermentation and create a hot sensation when imbibed) that are concentrated in the tails during distillation, and organic acids react to form more favorable fruity esters.

Chapter 17

BACKGROUND

———

BEER

As one of my favorite subjects to teach and to, ahem, research, beer is arguably the most important fermented product ever made. Think that's hyperbole? As covered in the previous chapter, there is actually a good bit of evidence for the fact that beer is the reason for human civilization. For now, though, let's cover the basics of beer and its production. Warning—while what follows may seem like an entire book, it is not. But, if you don't like beer, or just want to shy away from an entirely too long section written by the addled mind of an aging beer scientist, feel free to skip forward to a discussion of cider.

On the one hand, beer is incredibly simple. It's made from four basic ingredients—cereal grains, hops, yeast, and water. On the other hand, it's arguably the most complex fermented beverage. Cider and wine, for example, are made from two ingredients each. In the case of cider, that would be apples and yeast. For wine, grapes and yeast. The difference in complexity doesn't end there, however. Traditionally, yeast wouldn't have been added to apple or grape juice to initiate the fermentation. Rather, the yeast and bacteria present on the apple or grape skins would have been all the inoculant needed to get a batch going. Of course, we didn't know what microbes were until the mid- to late-nineteenth century, so it's not like brewers were purchasing their pitchable yeast from the lab to inoculate their batches either.

What they did know, however, was that they needed to transfer the krausen, the foam that forms on the top of a fermenting beer, from one batch to the next in order to initiate fermentation. Or if they were in Scandinavia, they may have used a magic stick in the form of a yeast log or a yeast ring, sometimes intricately carved logs or wreaths that had plenty of surface area on which to capture the yeast when these devices were dipped into a current fermentation before transferring it to the next batch. Again, the brewers had no idea what yeast was at the time. The mere notion of tiny, single-celled organisms was beyond anyone's comprehension. But they did know that the froth on top of a ferment, or the slurry that settled to the bottom of the ferment, had to be captured and transferred to the next batch to get it to ferment. The English even had a term for the magic stuff, *godisgoode*, which pretty much sums up their reverence for it. So, you could even argue that historically, at least, cider and wine production really required only one ingredient, with the necessary microbes coming along for the ride on the skins of the fruit.

The difference in complexity doesn't end there either. Most wines and ciders are made from one to three varieties of grapes or apples, respectively, per batch. Beer is often made with at least three different types of grains and as many types of hops per batch. That adds an exponential number of possible flavors that can be added to a beer relative to its fermented cousins. That's not even getting into the impact that water chemistry has on the flavor and mouthfeel of beer. The classic beer styles that we know and love were all born in different parts of the world based almost exclusively on the water chemistry of those specific regions and the types of beer that could be produced using that water.

Ironically, beer has been a victim of its own success. Because it was so important and consumed so regularly by huge swaths of the global population, it became known as the every-person drink. As a result, elitists, in their never-ending quest to separate themselves from the commoners, looked down their noses at beer. They needed a suitably snobby drink to replace beer, so who did they turn to? The French of course, who are revered worldwide for their ability to determine what foodstuffs, fashion, and art are suitably snobby or not. And what was their fermented beverage of choice? Wine! *Et voilà*—wine was elevated to a beverage that was suitable for the elites among us and beer was left to the commoners. If they only knew how much more complex beer was than wine, perhaps their roles would have been reversed.

Malting

So, how do we put those simple ingredients together to make beer? Well, first you start with the grain. The most common cereal grain used in beer production is barley, followed by wheat, rye, oats, and even corn for some styles. However, most of these grains cannot be added "raw." They must first be malted, a three-step process comprising steeping, germination, and kilning. Upon receipt of the grains from the farmer, the maltster (yup, that's what they're called) cleans the grains and sorts them according to size and fitness for use in brewing. Only the plumpest, highest quality grains go on to become beer. Then the steeping process involves soaking the grains in water, with periodic draining and air rests to clean them and allow them to respire, until they collectively reach a moisture content of about 46%. This moisture content kicks off the germination phase, the beginning stages of plant growth.

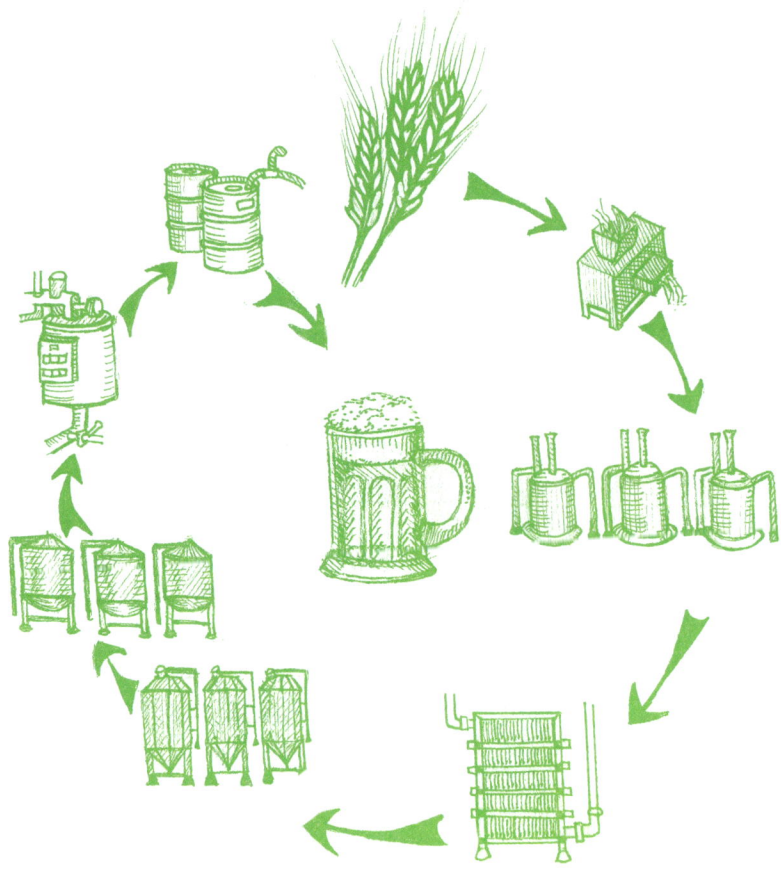

Plant growth hormones are synthesized in the embryo and sent to the aleurone, a small region of the grain that surrounds the starchy endosperm (the main portion of the grain that contains all the storage proteins and starch granules). The hormones signal to the aleurone that it's time to synthesize or activate the enzymes that will break down the material in the starchy endosperm to make it accessible for metabolism. First the cell-wall-degrading enzymes are synthesized. These break down the cell walls of the endosperm cells that hold all the starchy goodness. Then the protein-degrading enzymes are activated to break up the protein matrix in which the starchy granules are embedded. Next up are the starch-degrading enzymes. However, once those are synthesized, germination is stopped. If it were allowed to continue, those enzymes would break down the starch molecules into simple sugars that could be used as fuel by the grain that is trying to grow into a plant. Since we're trying to make beer and not more barley, germination is stopped, and the enzymes are maintained so that they can be reactivated in the brewhouse.

It is at this point that the final stage of malting, the kilning stage, is initiated. This does not just involve throwing the grains in a large oven and cranking up the heat. First of all, it depends on the type of malt that is being produced. The two main types are base malts and specialty malts. Base malts are more lightly kilned to maintain enzyme activity that can be used in the brewery. Specialty malts are generally more highly kilned to produce more color and flavor without a concern for maintaining enzyme activity since they are used only in small amounts in a brew. Either way, kilning starts off at a nice low temperature to remove most of the moisture in the grain and, if making base malts, to preserve the enzymes in the endosperm matrix. Then the temperature can be increased for different periods of time depending on the type of malt being produced. Ultimately, the moisture content should be reduced to about 4% by the end of kilning. After a storage period to allow the malt moisture content to equilibrate, the malts are ready for bagging and shipping to your local brewery.

Mashing

When the brewer is ready to use the malts, she must first mill the grains to the proper size. This depends on the type of mashing and separation system used in the brewery. I won't get into the details here, but the main systems available are separate mash and lauter tuns, a combination mash and lauter tun, and a mash filter. First of all, you're probably wondering what mashing and lautering are. Mashing is the process whereby the milled grains are steeped in hot water at a particular temperature to activate specific enzymes to produce the style of beer desired. Lautering is the process whereby the sweet wort, the sugary solution resulting from the mash, is separated from the remaining grain solids. So, without going into the details involving the Darcy equation that are a never-ending source of frustration to my students, a combination mash and lauter tun, the most commonly used system in a modern craft brewery, requires the largest grain particles; separate mash and lauter tuns, the older type of system still employed by many larger breweries, can use a finer grind; and a mash filter, only really employed by the largest breweries, can use a finer grind still, practically a flour.

Water

This is a good point at which to discuss the sometimes-overlooked ingredient in brewing, water, as this is the stage when most of the water that ends up as beer is added. Whenever discussing brewing water, I always think of the T-shirt that reads, "Save water, drink more beer." Cute, right? Unfortunately, it's also wildly inaccurate in its implication. Making beer actually requires quite a bit of water. In fact, the industry average, which includes the largest, most efficient breweries, is six volumes of water for every volume of beer produced. Many craft breweries probably greatly exceed that amount. Anyway, not only is a lot of water used to make beer but the quality of the water is also incredibly important. It's so important, in fact, that it is the reason why we have certain historical styles of beer. The incredibly high sulfate concentrations in Burton-on-Trent, England, which create a mineral bitterness, led to the first iterations of the hoppy, bitter pale ale and India pale ale. The crazy high carbonate concentrations in Dublin, Ireland, home to Guiness, provide plenty of buffering capacity for the more acidic, darkly roasted malts and led to the production of the dry Irish stout.

Nowadays, most breweries filter their water and often remove most salts using a reverse osmosis system so that they can start with a blank slate and add in the salts necessary to make the desired styles of beer. Do dissolved salts make that much of a difference in the finished products? You bet your sweet wort they do. One very current example is the difference between the West Coast IPA and the much maligned (by me) hazy IPA. The West Coast IPA is the natural evolution of the English IPA, a bitter (by English standards, although the original, pre–World Wars I and II version was quite bitter), but still quite malty and lower alcohol (by American standards, although, again, the original version was quite alcoholic) example of the style. The West Coast IPA is the grown-up, Americanized version of the more recent English IPA. It's often brilliantly clear; nice and dry, with lots of hop character and bitterness, a kiss of malt to almost balance the hops, and plenty of alcohol to provide that nice, warming sensation at the back of your mouth. If the West Coast IPA is the grown-up version of the English IPA, then the hazy IPA is the reversion back to the rebellious, goth teen. The hazy craze started with a couple breweries in Vermont (hence the synonymous New England IPA) making IPAs that were a diversion from the typical bitter bombs that defined the West Coast IPA. The two main differences between the styles boil down to the bitterness and the water chemistry, which I've summarized in table 17.1.

When you leave the differences at that, the hazy IPA is not even that terrible. Unfortunately, brewers saw the hazy beers that were coming out of Vermont at the time and failed to understand that the haze was primarily coming from hop residue from the ridiculous amounts of hops added late in the process. So, they came up with lots of ways to add haze (and instability) to their beers that I'm not going to go into here because, frankly, they trigger me. The moral of that story is that water chemistry matters. And if you're not careful with it, you may start an annoying trend that doesn't seem to want to go away anytime soon. Now back to the mash.

During the mash, the reactivated enzymes primarily degrade the starch molecules to fermentable sugars that can be metabolized by yeast. Most modern malts are sufficiently modified, which is the term used to describe the extent to which the enzymes have acted on the endosperm material during malting, such that the endosperm cell-wall material and protein

TABLE 17.1. Differences between West Coast and hazy IPAs

Characteristic	West Coast IPA	Hazy IPA
Hop additions	Early and often, leading to plenty of bitterness	Only late addition hops, leading to little bitterness but plenty of flavor and aroma
Water chemistry	2:1 sulfate to chloride ratio, providing a mineral bitterness to complement hop bitterness	2:1 chloride to sulfate ratio, leading to a sweeter, fuller mouthfeel
Taste and flavor	Bitter and delicious	Sweet and disgusting

matrix need no further degradation during the mash. Thus, unless certain under-modified malts are being used or particular styles of beer are being produced, generally one rest temperature (mash temperature) to activate the starch-degrading enzymes is all that is needed in the mash. So, once the starch is degraded, the lautering process can begin. The sweet wort is separated from the solids and run into the boil kettle as more hot water is added to the mash to wash as much of the extract from the grains as possible. This additional "wash" is referred to as the sparge. The spent grains are generally sold or traded to farmers and ranchers to feed livestock, a great way to make use of what would otherwise be waste material and support your local farmer at the same time. That said, there are other interesting options for using the spent grain material that will be explored more below.

Boiling

The boiling stage is what separates brewing beer from other fermented beverages and also from distilled spirits that otherwise share a similar production process up to this point. It's also one of the reasons that beer was safe to drink when the water would kill you. The alcohol content, low pH, and antimicrobial compounds from hops are other reasons as well, but the boiling certainly helps. You will sometimes see more reasons given, but I have boiled down (yes, I just went there) the main purposes of boiling to the following:

- halting enzyme activity

- sterilizing the wort

- concentrating the wort through evaporation

- developing color and flavor

- removing unwanted volatile flavor compounds

- achieving the required colloidal stability (you hear that, hazy IPA brewers?), and

- extracting bitterness from the hops.

HALTING ENZYME ACTIVITY AND STERILIZING THE WORT

Enzyme activity may have already been stopped if the temperature at the end of the mash was increased or if it was sufficiently increased during the sparge. If there are any remaining active enzymes prior to the boil, however, there won't be after the boil begins. The boil temperature is sufficient to deactivate if not denature every remaining active enzyme. The temperature is also sufficient to effectively sterilize the wort. The only possible survivors would be spore-forming bacteria. Bacteria in their spore form are pretty darn hard to kill, considering that some can even survive the harsh conditions of deep space. Even so, spore-forming bacteria should not be much of a concern in a brewery environment in the first place. And even if some were present and survived the boil, they wouldn't find a very hospitable environment in which to proliferate afterward, what with the low pH, antimicrobial compounds from hops, competition with other microbes for nutrients, and increasing alcohol content. So, we can consider the wort effectively sterilized after the boil.

CONCENTRATING THE WORT

Most breweries will achieve somewhere around a 10% evaporation rate per hour during the boil. Considering that most boils last for 1–2 hours, this concentrates the wort a good bit, causing the density of the solution to increase by a substantial amount. This increase in density is accounted for by the brewer, who presumably knows the evaporation rate and targets a particular density, the kettle gravity, prior to the boil, such that post boil, the correct density of the wort, the original gravity, is achieved. This is one of the more simple and important things that breweries do to achieve consistency in their products. The boil also facilitates one of the most ubiquitous and profound transformations that occurs whenever heat is added to food products, Maillard chemistry or Maillard browning.

DEVELOPING COLOR AND FLAVOR

There are a few browning processes related to foods. There is enzymatic browning, like what occurs when an apple is cut and left out to eventually brown. In this case, the particular enzyme found in apples speeds up the chemical reaction between the fruit tissue and oxygen in the air, leading to a browning. For nonenzymatic browning of foods (browning reactions that don't involve enzymes), there are effectively two processes, caramelization and Maillard browning. The difference between those two is in the starting materials. Caramelization involves the addition of heat to sugars and other carbohydrates. The heat (generally intense) causes molecular changes within the sugars, the products of which then start a long chain of complex reactions leading to all kinds of flavor and color molecules that we associate with caramel. Maillard chemistry involves heat and sugars as well but also an additional starting material: amino acids (the building blocks of proteins). Since almost all foodstuffs have sugars and amino acids, Maillard browning is by far the most common type of nonenzymatic browning. Throw your bread in the toaster to brown—Maillard browning. Your steak on the grill—Maillard browning. You get the picture. The amino acids may seem like a minor addition to the process, but for one they reduce the heat necessary for the reaction to occur. They also add in a world of complexity to the possible reactions and end products. All amino acids contain nitrogen and many contain sulfur as well. These elements are not present in sugars. Molecules containing nitrogen and sulfur are notoriously odorous, so the end products of Maillard browning, besides providing color, also provide a plethora of pungency. Now, imagine a boiling broth brimming with sugars and amino acids and the depth of flavor, aroma, and color products that arise.

REMOVING VOLATILE COMPOUNDS

Besides evaporating a good bit of water, volatile compounds in the boiling wort are also removed. Some may be volatiles that are not unfavorable, such as hop essential oils (more on that shortly), but the boil also removes unwanted compounds. The number one culprit is dimethyl sulfide (DMS). The aroma descriptors for DMS are plentiful, but the ones that capture its essence the best, in my opinion, are canned corn, tomato paste, or black olives (and not the good kalamata olives, I'm talking the nasty little ones that come out of a can and end up on way too many veggie pizzas). And it can be intense in high concentrations. I once had a bottle of it under the hood in my lab. When it was opened, even though it was under the hood, it would literally smell up the entire floor of our building. And that's a chemistry building that fully recycles the air every 12 minutes. We now have a bottle of it in one of our chemical fridges, the one that we affectionately refer to as the stinky fridge because, yup, it stinks up the entire facility every time the fridge is opened. Fortunately, if found in beer, it is in much lower concentrations and is reminiscent of canned corn, giving the beer a corny aroma and flavor. It is derived from the amino acid methionine in malt. During germination of the malt, a methyl group ($-CH_3$) is added to methionine, resulting in the unstable compound S-methylmethionine (SMM), which is the precursor to DMS. During kilning, much of the SMM, which is heat labile, degrades to the odoriferous DMS and is removed. However, if the malt being produced is a lighter base malt, the kilning process won't provide sufficient heat to degrade all the SMM. Thus, some will remain in the malt. So, the boil is the last chance to degrade the SMM and boil off the DMS. This is why it's so important to properly vent the boil, so that as much of the DMS as possible can be removed. However, for some lighter lagers that use very lightly kilned base malts, there will always be a little DMS that remains and that is acceptable for those styles. For all other styles, though, DMS is always considered an off-flavor.

ACHIEVING COLLOIDAL STABILITY

Colloidal stability is something that is done not only for aesthetic purposes but also to make the beer more stable from a chemical, physical, and microbiological standpoint. By removing solids that would otherwise remain in suspension, the potential for haze formation is significantly reduced. This greatly increases the shelf life of the product. But, hey, who am I to stand in the way of nonsensical trends that create a much less drinkable and shelf-stable product? During the boil, proteins in solution, primarily from the malt, denature and then begin to coagulate together much like the egg whites while cooking an egg. Polyphenols, a class of organic compounds that includes tannins and antioxidant compounds like curcumin and quercetin, extracted from both the husk of the malt and hops, will bind to the proteins and make even larger complexes. The larger the complex, the more likely it will be to settle out of solution. This can be further aided by the addition of kettle fining agents. These are generally derived from the red algae Irish Moss, *Chondrus crispus*, which contains the active ingredient κ-carageenan, a negatively charged polymer that binds to the proteins in solution, which are largely positively charged at the pH of the boiling wort. Once the boil is finished, wort is whirlpooled for roughly 15–30 minutes and allowed to rest for an equal amount of time. The whirlpool takes advantage of the same physical principles that are used when a cup of tea with loose tea leaves is stirred to focus the tea leaves at the bottom of the cup: namely, the centrifugal force pushes the solids to the edge of the kettle, and the frictional forces drag the solids to the bottom middle of the kettle. What is left is referred to as the hot break, a mix of coagulated proteins, polyphenols, fining agent, and hop material that is hopefully left behind in the kettle when the wort is transferred to a fermentation vessel. If it is not left behind, the material can lead to a number of issues, including possible microbial infections, off-flavors, colloidal instability in the finished product, and worst of all, hazy IPAs.

EXTRACTING BITTERNESS FROM THE HOPS

You may not know it based on my railing against hazy IPAs, but I love hops. Just not in my orange juice, smoothies, or milkshakes, which is what most hazy IPAs remind me of. *Humulus lupulus*, hops, is a perennial bine native to Europe, Asia, and North America, but which is cultivated on every continent now except for Antarctica (although with climate change, we may not have too much longer to wait for that one as well). It spreads via rhizome and grows as much as 10 meters in a season. It's a dioecious plant with male flowering plants and female flowering plants. The females, however, are the ones with the goods, and the males are allowed to live only for breeding purposes. Hops are in the *Cannabinaceae* family, and its closest botanical cousin is cannabis. Once you see a hop flower or cone (strobilus if you want to be technically correct) and smell the essential oils, you will recognize the family relation. In fact, hops and cannabis share many of the same essential oils. These essential oils provide flavor and aroma to beer in the form of floral, fruity, spicy, and earthy notes. The other important compounds in hops are the bittering acids, which provide the bitterness to beer. Without bitterness, beer would be cloying from the sweetness of the malt.

Historically, all kinds of compounds were used to provide bitterness to beer, some of which, like wormwood, are potentially toxic. It took a few thousand years to settle on hops as the main bittering agent in beer. In England, beers were even distinguished according to whether they were hopped or not. Ales were unhopped, while beers were made with hops. King Henry VIII was apparently not a hophead. He famously declared hops a wicked and pernicious weed, which is where Wicked Weed in Asheville, North Carolina, got its name.

The two main bittering acids are alpha acids and beta acids. They both undergo chemical transformations prior to providing the bitterness, though. Beta acids oxidize during the boil, and the oxidized beta acids are responsible for about 10% of the overall bitterness. Alpha acids have to isomerize, or rearrange some chemical bonds to change their molecular shape, so that they become soluble in wort. The isomerized alpha acids are responsible for roughly 90% of the overall bitterness in beer.

The length of time that the hops are boiled determines their ultimate impact on the beer. When hops are added at the beginning of the boil, the bittering acids are extracted and have a chance to isomerize and oxidize, thereby providing bitterness. However, any essential oils added at this point readily boil off since they're so volatile. When hops are added toward the end of the boil or even after the heat has been turned off prior to whirl-pooling, more of the essential oils remain, adding their flavor and aroma to the wort. At this point, though, very little bitterness is extracted. Hops are also often added during or after fermentation, a process referred to as dry hopping, which makes no sense since the hops get quite wet after adding them to the fermenter.

Aeration

Once the heat is turned off in the kettle, the wort is whirlpooled and allowed to settle as discussed above. Following the whirlpool, the wort is rapidly chilled, typically using plate heat exchangers, and subsequently oxygenated on its way to the fermentation vessel. Oxygen is the enemy of beer, leading to staling reactions and all kinds of off-flavors. So, why is oxygen added to the chilled wort prior to fermentation? Well, since you asked. The yeast must reproduce to make enough cells to ferment the wort. Yeast reproduces asexually through budding (they can sexually hybridize as well, but that's a story for my upcoming romance novel, *Beauty and the Yeast*). To do so, the yeast cells must produce all the material for new cells, including cell membranes, which require sterols and fatty acids. Fats are not welcomed in beer production, as they can oxidize and produce compounds that lead to stale off-flavors. So, without fats added to the wort, the yeast cells must make the fat themselves. Which they can. But that requires oxygen. Not to worry, though, if the correct amount of oxygen is added to the wort, the yeast will mop it up within 24 hours and make lots of new daughter cells with it. And that is the only time that oxygen should be introduced to the process.

Fermentation

Once the wort is chilled, aerated, and added to the fermenter, the yeast can be pitched. There are two main species of yeast that are used to make beer, *Saccharomyces cerevisiae*, ale yeast, and *Saccharomyces pastorianus*, lager yeast. Within each of the two species, though, there are dozens if not hundreds of different strains. *S. cerevisiae* is a mesophilic (enjoys moderate temperature ranges) yeast that is the most widely used microbe in industry. It's the same species used to make most bread and wine, for example. Most *S. cerevisiae* strains prefer fermentation temperatures in the 18–22°C (64–72°American) range. It was historically referred to as a top fermenting yeast due to the higher fermentation temperatures, which caused a vigorous ferment resembling a boiling liquid with lots of carbon dioxide bubbles that carried the yeast to the top of the solution. The foam that formed on top of the solution as a result, the krausen, held a good bit of the yeast, which was then top-cropped to pitch into the next batch. The warmer fermentation temperatures of *S. cerevisiae* are used to produce ales, which are characterized largely by the flavor profile provided by the yeast itself. Ale yeast, encouraged by the higher temperatures, produces more metabolic byproducts, namely esters, which are responsible for many of the fruity flavors associated with ales. Higher fermentation temperatures mean the ferment goes faster too, which is why the vast majority of craft beer produced is ales, as the quicker turnaround time means higher throughput and more beer. *S. pastorianus* by contrast, is a more cryophilic yeast, preferring it a bit chillier. The preferred fermentation range for this species is 8–12°C (46–54°American). The cooler fermentation temperatures result in a less vigorous ferment, so that historically lager yeast was referred to as bottom fermenting. Fewer carbon dioxide bubbles during the ferment meant less krausen on top, and the yeast would largely hang out at the bottom of the fermentation vessel. Cooler temperatures also lead to a "cleaner" ferment, with fewer metabolic byproducts produced by the yeast, which is also less genetically inclined to produce as many esters. As a result, lagers have very little of their flavor profile contributed by the yeast. This lets the malt and hops shine through. However, the cooler fermentation temperatures also mean a longer ferment and lower throughput in the brewery. This has led many craft breweries to shy away from doing lagers, as they take too long to be financially tenable for a small brewery. When it comes to mass-produced beer and overall volume of beer produced, though, the opposite is true. Lagers dominate the beer world. They are by far the most popular beers in the world by volume produced.

For a story of international fungal affection, I've provided an excerpt from my yeast romance novel, *Beauty and the Yeast*, below.

EXCERPT FROM *BEAUTY AND THE YEAST*

It was the fifteenth century in rural Bavaria, Germany. Johann the yeast cell was from a long, proud line of *Saccharomyces cerevisiae* yeast that had been fermenting German ales for weeks, if not months. So, to say that Johann was proud to be a top fermenting German yeast cell would be an understatement. But he wasn't naive to the current politics either. He had heard of the rumblings from the Duke's court about some big changes that could impact him and all his daughter cells. Only cold weather brewing? What in the schnitzel was up with that? I mean, sure, there had been some instances of bacterial invasions on the southern front that led to infected and soured beers, but was that reason enough to change hundreds of years of proud brewing tradition? And what would that mean for him and all his progeny? Dormancy, if not death and lysing, is what it meant.

He was so caught up in his own thoughts that he barely noticed when the ravishing stranger floated into view. But Inez, one of the recent *Saccharomyces eubayanus* newcomers to arrive in the fermentation barrel, certainly noticed Johann's bulging organelles and spherical perfection, with a few bud scars that seemed strategically placed for maximum attractiveness on his otherwise unmarred and pleasantly fuzzy cell-wall surface. She floated closer to him and, before bumping into him, she asked in her most innocently alluring voice, "Excuse me, but do you know if there will be more maltose available soon? My daughters and I haven't eaten for what seems like minutes."

When he heard that smoky, sultry voice, Johann certainly came to attention. "I'm sorry, miss . . . ?"

"Inez. I asked if there was more maltose available for my daughters and me. We just arrived. And you are?"

"Pleasure to meet you, Inez. I am Johann. I believe all the fermentables have been added already. But you are welcome to my meager stash. By the way, where are you coming from? I can't quite place that accent."

"Well, that's a long story. I can only remember a couple generations back, but there are traces of memories from Ireland, Asia, and even a faraway land called Patagonia. I believe that we ended up in this fermenter accidentally, though, from a stray beard hair."

"Yah, well that happens sometimes. I always say the brewers should trim their beards, but you know brewers and their affinity for facial hair. By the way, is it cold in here? I barely have the motivation to eat when it's this cold."

"Cold? This is downright warm in here, if you ask me. And I'm famished. I could eat all day."

Not cold? In these conditions? Well, that sealed the deal. Johann knew in that instant that his destiny was to sexually hybridize with this bewitching newcomer. He knew the time was up for him and his kind. If he was to have any say in the future of German brewing, his only chance was by making beautiful daughter cells together with Inez.

"Inez, we've known each other for quite some time, now. I mean, at least two minutes at this point. And I know that this may seem rather forward, but would you like to shmoo with me?"

"I thought you would never ask."

The next couple of hours were a blur of bulging nodules and combining DNA. It was wet and wild, fueled by desire and the alcohol in the fermenter. Was that Hans Sachs playing in the background too, or just the rhythm of their organ-

elles beating in harmony? It didn't matter. Nothing mat-
tered for that brief but seemingly endless period of bliss.

And then it was over. But the result of their romantic in-
terlude, *Saccharomyces pastorianus*, would go on to change
the brewing world. It turned out that Johann was right, after
all. The Duke dictated that beer could be brewed only in the
cold weather months. This edict was later passed into law.
The time for *Saccharomyces cerevisiae* in Germany, except
for in a few areas, had come to an end. But Johann and Inez
were able to live on together through their beautiful, cold-
tolerant progeny.

I hope you enjoyed that excerpt from my upcoming tale of tiny yet titil-
lating temptation. In case you didn't get the gist of it, for which I would
certainly not blame you, it's supposed to cover the rare hybridization event
between *Saccharomyces cerevisiae* and *Saccharomyces eubayanus* that led
to *Saccharomyces pastorianus*, lager yeast. The lineage of *S. pastorianus*
had been a mystery for many years. It was clear, based on DNA evidence,
that *S. cerevisiae* was one of its parents (the father in my fictional tale, although yeasts don't have a gender per se). The other parent was hypothe-
sized to be *Saccharomyces bayanus*, a fairly common yeast used in wine,
cider, and the production of distilled spirits. However, the DNA wasn't a
perfect match, so it was evident that *S. bayanus* was probably more like
a drunk uncle, a close relation that you tolerate at Thanksgiving, but not
the real parent. The hunt for the derelict parent went on for years. All the
big guns were called in—Dr. Phil, Jerry Springer, Maury—but to no avail.
Until, that is, a group from Argentina, while searching for likely candidates
in *Nothofagus* (southern beech) forests in Patagonia in 2011, tracked down
the deadbeat parent on sugar-filled galls common to those trees. The newly
discovered species was named *Saccharomyces eubayanus* because of its
genetic similarity to *Saccharomyces bayanus* (Libkind et al. 2011). But the
question remained as to how the estranged parent made it from the southern tip of Argentina to Germany in the fifteenth century or earlier, when
lager beers likely first started to appear there. Since the original discovery
of *S. eubayanus* in Patagonia, the species has also been discovered in Asia,
North America, and even Europe. That said, based on the genetic composition of *S. pastorianus*, modern lager yeast, and the *S. eubayanus* strains dis-

covered around the world, the strains from Patagonia seem to be the most likely source for the more recently discovered strains in other parts of the world as well as for the hybridization with *S. cerevisiae* that led to the lager species. So, the mystery remains regarding the relocation of our randy, cold tolerant yeast species. Was it the Vikings after all? Migrating birds? Aliens? Who knows if we'll ever find the answer. But to whomever was responsible for the transport, I say thank you for all the delicious lagers!

During fermentation, the yeasts consume the fermentable sugars in solution and produce ethanol and carbon dioxide through the alcoholic fermentation metabolic pathway. This is a unique evolutionary adaptation to rid the cell of toxic intermediate byproducts, specifically pyruvic acid, produced during glycolysis, the initial metabolic pathway employed to extract the available energy from the sugars that were consumed. Brewing yeasts are facultative anaerobes, meaning they are capable of aerobic respiration in the presence of oxygen. This is a metabolic pathway, an alternative to alcoholic fermentation, that extracts far more energy per sugar molecule. That's why aerobic organisms, like ourselves, regularly employ it. However, even in the presence of oxygen, if there is just a little bit of glucose in solution, the yeast will revert to alcoholic fermentation over aerobic respiration. Choosing to make ethanol at the expense of additional energy extraction? I sure do love yeast!

Yeasts also take up nitrogen from solution to make the necessary proteins. That, along with the byproducts from lipid metabolism after oxygen uptake, results in the production of lots of flavor active compounds. The production of these compounds depends on environmental factors, available nutrients, and to a large degree, the genetic predisposition of the yeast strain. Ethanol doesn't have much flavor, and carbon dioxide just adds a little acidity and fizz, so without the production of these metabolic byproducts, the flavors would come entirely from the malt, hops, and brewing water. This is why lagers taste so "clean." Lager yeasts typically produce far fewer of these byproducts relative to ale yeasts under normal fermentation conditions, which generally produce them in abundance.

Maturation

Ale fermentations generally take between 3 and 7 days, whereas lager fermentations usually take around 2 weeks because of the cooler temperatures they require. After this primary fermentation, there is a maturation period. For ales, this is pretty much to ensure proper clarity (especially if not filtering the product) and the absence of detectable diacetyl, a small organic molecule that has the flavor and aroma of butter or butterscotch candies (it is the compound that gives butter its flavor and that is used in artificially flavored "buttered" popcorn). All yeasts indirectly produce diacetyl during fermentation, but they will also absorb it and degrade it to a flavorless compound if they are healthy and given adequate time. Many ale yeast strains that were originally isolated in the United Kingdom are notoriously flocculant, meaning they clump together and settle out of solution very efficiently. So efficiently, in fact, that they end up leaving behind a good bit of diacetyl. This, I have no doubt, is where J.K. Rowling got her idea for butter beer in Harry Potter. Since so many traditional beers from this region have detectable levels of diacetyl, it is actually acceptable for many of the styles born in this region. Other than those styles, though, it is always considered an off-flavor. Unless you're a student at Hogwarts, that is.

Lagers actually undergo a secondary fermentation, which is the lagering phase. Lager roughly translates as "store," as in "to store" something, like your beer. Traditionally, secondary fermentation would last as long as 2 months at close to freezing temperatures. The lager yeast would continue to slowly ferment the solution and ultimately mop up the diacetyl and even carbonate the beer after the fermenter was closed off, a process referred to as spunding. Nowadays, modern techniques and hybridized strains allow most breweries to lager their beer in 2 weeks or less. This makes it much more manageable for smaller breweries to produce lagers, which is why there are many more craft lagers available today relative to a few years ago.

After maturation, beers can be clarified, if necessary, through the addition of fining (clarifying) agents, filtration, or centrifugation. Then they are carbonated (if they weren't already) and packaged. Larger breweries will typically pasteurize their beers as well, which helps ensure shelf stability. At this point, they're ready to be enjoyed.

CIDER

After slogging through that last section, you may never want to drink beer again. Well, good thing there's cider! Cider popularity in this country has soared over the last decade. It is certainly delicious and refreshing, but I believe that a lot of that growth is the result of people shying away from gluten-containing foods and beverages, in some cases for very good reasons, like celiac disease, and in other cases for merely perceived intolerances. Regardless, it created a huge market for gluten-free foods and beverages.

Since most beers have some gluten in them, people sought an alternative alcoholic beverage. In all my prognosticative wisdom, I was sure that the alternative, gluten-free beverage would be mead. Alas, I was sadly mistaken. Well, only sad because I was dead wrong. The fact that cider filled that gap admirably made me perfectly happy, as there are few things in this world as refreshing and delicious as a good, dry hard cider on a warm, sunny day. Or on a cold day, sitting by the fire. Or on a moderate-temperature, rainy day. Cider is one of the easiest fermented beverages in the world to make, but it also takes a lot of practice and attention to detail to make it well, starting with the ingredients.

Apples

First and foremost, not surprisingly since they're the only ingredient besides the yeast and possibly the bacteria used to ferment them, is selecting the right apples for the intended cider. No apple is bad per se, but there are certain characteristics to consider for cider apples. Sugar, acid, pH, aroma and flavor, and nitrogen are all characteristics that will ultimately determine the fermentation performance and quality of the final product.

SUGAR

Sugar content in apples generally ranges from 6% to 20%, with most commercially available apples falling in the 8–12% range. This is an important consideration for cider not only from a product formulation perspective but also from a legal one. The Alcohol and Tobacco Tax and Trade Bureau (TTB), a federal government agency, collects taxes on all legally produced alcoholic beverages that are intended for sale. For cider that is at least 7% ABV (up to 24% ABV—yowzah!); the product is considered "apple wine," and the TTB handles the taxation as well as the labeling requirements. For a cider that is under 7% ABV, the product is considered hard cider and is taxed by the TTB, but in these cases, the labeling is handled by the FDA rather than the TTB. That's why most craft ciders in this country are at least 7% ABV, since the product can be produced with a wine license and dealt with like any other product. Most craft cideries choose not to deal with two federal agencies, as dealing with one is generally enough to keep any producer on her toes. So, what does all this mean with respect to the sugar content of apples? Roughly half the sugar in a ferment is converted to ethanol, depending on the yeast strain and fermentation conditions. So, if the apples you're using to make cider have an average sugar content of 12%, that means, without any additional sugar, your cider will likely be around

6% ABV and subject to both TTB and FDA regulations. So, yeah, sugar content matters.

ACID AND pH

The acid content of apples plays a role not only with the pH of the cider but also with the final flavor. Apples' acidity comes primarily from malic acid, which determines the sourness of the apple and the resulting cider, but also from the much lesser contributions of galacturonic, citric, quinic, and formic acids. (Formic acid, incidentally, is the acid produced by many ant species that cause the pain when they bite or spray!) Acid concentration, or acidity, and pH are related, but the acid concentration of apples does not directly translate to pH. The pH is a measure of the concentration of hydrogen ions in solution. The greater the number of free hydrogen ions in solution, the lower the pH will be. Acidity is a measurement of the acid concentration (e.g., the concentration of malic acid in an apple) in solution. Acids can exist in two forms, with a bound hydrogen atom (the neutral state), or in the ionized form if the hydrogen is released as a hydrogen ion. If there are more organic acids in apples and, hence, the cider, then there is greater potential for hydrogen ions to be released into solution. However, that depends on the existing pH and buffering capacity of the solution.

Don't sweat the details. The important take-home is that a higher acid concentration generally means a lower pH, but acidity and pH are separate measurements, and one cannot be directly determined by the other without additional information. The sourness of an apple is not directly related to the pH but rather a result of the amount and types of acids in the apple. The pH plays a bigger role in microbial stability and the efficacy of sulfur dioxide, which is used as an antimicrobial agent. A lower pH inhibits most pathogenic bacteria directly and also makes sulfur dioxide much more effective in achieving the same goal. Generally, you want to keep the pH of your cider below a maximum of 4.0, but 3.6 or less is ideal.

AROMA AND FLAVOR

Whereas the sugar and acid content of cider can be adjusted before or after fermentation, the flavor and aroma profile of the apples cannot be adjusted. This is why it's critical to find apples, either a single varietal or, more commonly, a blend of apple varieties that have a flavor and aroma profile that is pleasing to the consumer. This can be challenging, as the same apple varieties will have different flavor and aroma profiles, depending on where and how they're grown. Also, you may love certain varietals, but they might be unavailable in your location. Apples are bulky, heavy, and perishable, so it's difficult and expensive to ship large quantities very far. Typically, the apples used in cider will be sourced from a fairly close locale. So, you're generally dependent on the varietals that are grown close to you.

YEAST ASSIMILABLE NITROGEN (YAN)

Just as you and I need proper nutrition to survive and thrive, so too do yeast cells. And nitrogen is one of the most critical nutrients for yeast cells (and humans for that matter). Nitrogen is one of the key elements in amino acids, which make up proteins. Yeasts (and humans) need proteins, as they are the workhorses of pretty much every physiological function. To ensure they have the adequate amounts and types of proteins to function properly, yeasts must have adequate amounts of yeast assimilable nitrogen (YAN) in solution. YAN comprises ammonium ions (NH_4^+) and free amino nitrogen (FAN, or amino acids that can be used by the yeast), both of which are the forms of nitrogen that can be taken up by the yeast and used to make proteins. If the apple juice is deficient in YAN, this can result in a sluggish or stuck fermentation as well as off aromas and flavors, the result of reduced sulfur compounds, primarily hydrogen sulfide, which is the aroma of rotten eggs and, well, farts.

A cider ferment should not take weeks to complete. If it does, it is likely that the juice is deficient in YAN. Apples, and most fruits for that matter, generally lack sufficient amounts of YAN for a healthy fermentation. The general guideline is a minimum of about 10 milligrams of YAN per liter of apple juice for every percent sugar to be fermented. It is obviously best if you can measure YAN (typically done as a sum of separate measurements of FAN and ammonia nitrogen) and add in the correct amount to account for any deficiency. This is not something that can generally be done by the home cidermaker, however. Regardless, adding in a recommended dosage of nutrient will do the trick since apples are generally pretty low in nitrogen, so you won't run the risk of having too much nitrogen in solution.

There are two main sources of nitrogen rich yeast nutrient—diammonium phosphate (DAP) and yeast autolysate-based nutrient. DAP is a source of ammonia nitrogen that is readily taken up by the yeast. It's inexpensive and easy to use, but you don't want to use this as your only source of nitrogen, as yeasts want some ready-made amino acids as well. This is where the yeast autolysate-based nutrient comes into play. Yeast autolysate is a fancy term for dead yeast cells that have spilled their guts, which have lots of amino acid nitrogen, vitamins, minerals, sterols (for building the cell membranes of new cells during yeast growth), etc. The nutrients are made from healthy yeast cells that have been fattened specifically for this purpose, not unlike a Thanksgiving turkey that has been fattened to become the centerpiece of that meal. There are many options for this type of nutrient, including Fermaid K, Fermaid O, and Servomyces. You could use these nutrients as the sole source of nitrogen, particularly if they also include DAP, many of which do, or you could use a 50/50 blend of DAP and yeast autolysate nutrient. You can add all the nutrient at the beginning of fermentation, but many cidermakers prefer to add half at the beginning of fermentation and the other half after about one-third of the sugars in the ferment have been consumed.

Pressing and Clarifying

Now that you've got the proper apples selected for your cider, it's time to process those apples into juice. To do so, first you have to grind or pulverize them in some manner. Typically, particularly for smaller producers, this is done with a grinder (which can be an actual apple grinder or even a garbage disposal—been there, done that), followed by a press to squeeze the juice from the ground apples. Presses come in many different types and sizes, so make sure to choose one that is an appropriate size for the job. For an estimate of the amount of juice you can expect from apples, a good guideline is about 55 to 70 gallons of juice for every 1,000 pounds of apples, depending on the apples, the press, how the process is treated, and so on. To aid in the yield of juice, maceration enzymes can be added to the pulverized apples in the press. These are typically not only a combination of enzymes that help to break down the solids and increase the yield of juice but also include pectinase to help clarify the juice before the next step, fermentation. These enzymes are not necessarily cheap, though, so choose wisely and use them only according to the specifications.

Once the apples have been pressed into juice, the juice must be clarified to remove solids like skins, seeds, pulp, debris, insects, and suspended solids. This helps not only to clarify the juice and ultimately the cider but also to reduce the spoilage potential as well as the potential for off-aromas like the farty, sulfur aromas mentioned earlier. Generally, the juice can be left in a settling tank overnight to allow it to settle before transferring the juice off the settled solids and into a fermenter.

A dose of sulfur dioxide is typically added to the freshly pressed juice as well not only to reduce the potential for unwanted microbes taking hold and kicking off a spontaneous fermentation but also to reduce the oxidation potential of the juice and to preserve its fresh character and prevent brown-ing. A typical dosage of sulfur dioxide (added as potassium metabisulfite, or sometimes sodium metabisulfite, both of which convert to sulfur dioxide in solution) is 30–50 mg per liter of juice. Sulfur dioxide binds to many com-pounds in solution, so by the next day when the juice is settled and ready for fermentation, very little free (unbound) sulfur dioxide remains in solu-tion to prevent fermentation. This is also the point at which YAN would be measured (or not) and nitrogen added, and the pH, acidity, and sugar levels adjusted to the desired ranges. Now the juice is ready for fermentation.

Fermentation

In order to ferment the juice into hard cider, the juice is inoculated with yeast. There are lots of yeast species and strains that will do the job, but to make the cider with the desired specifications, the right yeast must be selected for the job. The main characteristics to be considered are the al-cohol tolerance of the yeast, sulfur production (i.e., how farty will it be), the preferred fermentation temperature, and the flavor and aroma profile that will be produced. Yeast can have a huge influence on the final flavor profile of the cider, or a yeast strain can be selected that is rather neutral, one that won't produce much flavor and will allow the flavor of the apples to shine through. It just depends on what you want in your final product. The key, once the yeast strain is selected, is to use the healthiest yeast possible, whether it's a liquid culture or dried yeast, and to pitch the correct amount of yeast for the job. This will help to ensure a healthy ferment without the threat of spoilage microbes competing for nutrients.

Once fermentation is complete, the cider should be assessed for residual sugar content, acidity, clarity, and overall flavor and aroma. The sugar can always be increased if the cider is too dry for your liking. You could blend in some sweet cider or add in apple juice concentrate, table sugar or honey, but that could lead to refermentation. Another option would be to add sorbitol, which is a naturally occurring sugar alcohol that is found in apples already, so you're not adding anything "unnatural" or "foreign" to your cider. The benefit of sorbitol is that it is not fermentable by most microbes (some bac-teria can ferment it, but it's not common).

Another way to ensure there is some residual sugar in cider, if your aim is to make a semisweet or sweet cider, is to arrest fermentation prior to completion. There are several ways to do this. One is to cold crash the fer-menter to temperatures below the yeast's tolerance. However, there is still the possibility, as there is residual sugar in the cider, that another more cryotolerant microbe that is present, unbeknownst to you, may see this as their opportunity and begin munching on the sugar. To prevent this, sterile filtration, which filters out all microbes, or pasteurization is always an option, but not typically for the home cidermaker.

Otherwise, additional sulfur dioxide should be added, as it should be at the end of fermentation regardless; the amount should be adjusted to the correct level to protect the cider from refermentation. What is this amount, you ask? Well, that depends on the pH of the cider and how much becomes bound to other compounds and cannot do its antimicrobial work. Ultimately, though, somewhere around 50 mg of free sulfur dioxide per liter of cider should be plenty effective and still well below the legal limits for wine. I wish I could make it simpler than that, but accurately measuring the amount of sulfur dioxide in your cider is not something that you're going to be able to easily do at home, which, unfortunately, precludes you from knowing exactly how much to add. All is not lost, however. There are plenty of effervescent sulfur dioxide tablets that can be used by the home cider- or winemaker, with recommended doses that should at least get you in the ballpark.

The acidity can be adjusted as well, but this is slightly more complicated for ciders. The main organic acid in apples is malic acid. Some lactic acid bacteria, *Oenococcus oeni* to be specific, can happily convert malic acid, which has a very sharp acidity, to lactic acid, which has a much softer acidity, in a process called . . . wait for it . . . malolactic fermentation. If your cider has an unpleasantly sharp acidity, you may want to consider malolactic fermentation. If you're not careful about your sanitation, it could happen on its own if the right bacteria are present, often together with other unwanted off-flavors as well. Otherwise, you can inoculate your cider at this point with a commercial strain, which will give you much more consistent results. To increase the acidity in your cider, you can add acid as well. Tartaric acid is preferred for this over other possible options like citric or malic acids.

Clarification

Clarifying ciders is done not only for aesthetic purposes but also to stabilize them and prevent any unwanted microbial activity. By removing solids, excess proteins and carbohydrates that undesired microbes could feast on are eliminated, but also many of the microbes themselves are removed, albeit not necessarily completely, as they can be part of the suspended solids. The simplest way to clarify cider is to cool it down, that is, cold crash it, and let the solids naturally settle out. Then the cider can be racked off the sediment (transferred) into another tank. The key to cold crashing is to do it slowly, even though this seems to contradict the term used to describe the process. This keeps the yeast viable and able to finish up its job (if there's anything left to do). If the ferment is completely finished and you have no plans to reuse the yeast, then you can cold crash in a manner more fitting the name—as in fast. This will shock and potentially kill most of the yeast, but if you're not using them again and rack your cider off them before they have a chance to lyse and spill their guts, go for it.

Cold crashing alone may not do the trick, though. In that case, enzymes and filtration may be necessary. Enzymes can help to degrade pectins and proteins that may have partially precipitated during cold crashing to create a colloidal haze. There are many options for filtration depending on whether you just need a coarse polishing or a full sterile filtration. Regardless, this is not something that would normally be accessible to the home cidermaker. If not, fining agents that can be added to remove different solids through both chemical and physical means could be employed. There are plenty of options out there, generally available anywhere that you can get cidermaking equipment. It's always better to know what type of suspended solids are causing the haze so that the most appropriate fining agent can be selected. Many fining agents are compounds that carry either a positive or a negative charge and will chemically bind to oppositely charged particles in the cider. The larger the flocs that are produced through this process, the more likely they are to settle out of solution. Once settled, the cider can be racked off the sediment for a nice, clarified product. Again, though, it may not be possible to determine the composition of the solids. Fortunately, many of the fining agents work through mechanical or physical action as well, by creating a three-dimensional web-like network that settles out of solution and pulls any solids out with it as they get caught in the web. Once your cider is clarified, it's time to chill (literally and figuratively), carbonate, and enjoy!

WINE

Wine production is fairly similar to cider production, which makes sense if you think about it. They're both fermented fruit juices when you get right down to it. Some of the main differences lie in the way the fruit is processed, and white and red wines differ slightly in the ways they are made. Obviously, it all starts with the grapes. As the saying goes, you can make bad wine from good grapes, but you cannot make good wine from bad grapes. So, make sure that you are sourcing your grapes from a reputable vineyard, whether it's a commercial vineyard or one that's in your backyard. If you are growing your own grapes, take the time to learn how to do it properly. There are plenty of materials, workshops, and even classes on this offered through most local agricultural extension offices.

There are a number of factors that influence the overall quality and flavor of the grapes, qualities that will ultimately be transferred to the wine. These include the local climate, soil, vineyard location, pests and diseases, vineyard practices, and the harvest date. The maturity of the grapes, or chemical and physiological ripeness, is critical to determining the proper time to harvest. Chemical ripeness refers to the concentration of sugars, acids, and the pH of the grapes. If grapes are harvested too early, the sugars will be too low and the acids will be too high. Unless you're making a sparkling wine, this is generally not desirable. Additionally, the physiological or flavor ripeness refers to whether the seeds are ripe and if the proper flavors for that grape varietal have developed. Keep in mind also that any diseased fruit will impact the overall quality of the wine, so make sure that the grapes you use are free of molds or other signs of disease. Sometimes, however, despite your best intentions, the weather just ends up getting in the way. If bad weather is forecast for harvest time, you may need to harvest a bit earlier than intended, lest the rain spoil the whole crop.

Pressing

Once the grapes are harvested, they must be processed in fairly short order. Grapes are quite perishable, especially when you've got a ton of them smushing down on themselves in a bin. And if the juice begins to ooze out of the grapes, they will begin fermenting if you're not careful. The skins already have yeast and bacteria on them, plus you will attract pretty much every yellowjacket in the area (they seem to love grapes!), and yellowjackets are also great vectors for yeast and bacteria. So, once the grapes are harvested, it's time for destemming and possibly crushing and pressing. This is where the process differs for white and red wines.

For white wines, the grapes are destemmed and the berries are added to a press. The grape must, which includes the grapes and juice, is squeezed in the press to separate the juice from the skins and seeds of the grapes. The juice is then collected in a tank where it is clarified either mechanically with a centrifuge or through settling, a process also referred to as débourbage. For rosé wines, which are made from red grapes, the grapes are destemmed and crushed, and then moved to a vessel where the juice remains in contact with the skins for at least a couple of hours and up to a couple of days until the desired color and flavor has been extracted from the skins (for most red grapes, the color—and tannins—is exclusively in the skins, whereas the interior of the grape is light colored like white grapes). At this point, the grape must is pressed, as with white grapes, to separate the juice from the seeds and skins and mechanically clarified or allowed to settle. With red wines, the grapes are destemmed and crushed and then moved to a fermentation vessel. The juice, skins, and seeds are fermented together so that color, flavor, and tannins can be extracted from the skins (and seeds to a degree).

Clarification

The pre-fermentation settling or clarification for white and rosé wines is done to remove solids that can include the yeast or bacterial cells that were hitching a ride on the grapes, tartrate salts, proteins, pectins, tannins, and pieces of grape skin, pulp, seeds, or stems that made it through the press. Not only does this clarify the juice but it also helps produce fresher, fruitier wines with better color; helps prevent an unintended wild fermentation; reduces sulfurous off-flavors; and improves fermentation characteristics, since the inoculated yeast don't get stuck under sediment as they're trying to ferment. If the juice is allowed to settle without any mechanical clarification, the temperature is reduced to 5–10°C (41–50°American), which helps the particles to settle and reduces potential microbial activity. Clarification or settling aids such as pectinase enzymes, bentonite or other fining agents, can be employed to help improve the settling velocity and efficiency.

Once the sediment, referred to as lees or, more specifically at this stage, gross lees (because they're gross—no, the larger particles at this stage of the process are referred to as gross lees, whereas the finer sediment that settles out later in the process, made up of smaller particles, is referred to as fine lees), settles on the bottom of the tank, the clear juice can be transferred off the lees, a process referred to as racking. This is generally done within 24–48 hours after pressing. For larger wineries, the lees can then be filtered to remove any residual juice from the solids. If a centrifuge is employed to clarify the juice before fermentation, this generally achieves a much more thorough separation of the solids from the juice, so no further filtration of the lees is required.

Another mechanical clarification option, which is certainly more cost-effective than centrifugation (at least until that centrifuge is paid off—they ain't cheap), is flotation of the pressed juice. Nitrogen gas is pumped into the tank to saturate the juice in the tank. Nitrogen gas is not very soluble in water, so the bubbles cling to the solids suspended in the juice and float them to the top of the tank. A foam cap forms on top of the juice and can then be separated from the clarified juice below. There are several benefits to this clarification method besides being inexpensive: it can be done 4–8 hours after pressing; it is very effective (no lees filtration required); and can be done at fermentation temperatures (14–16°C / 57–61°American), which saves time and energy that would have had to be spent warming the juice to fermentation temperatures.

Fermentation

Now that the juice has been clarified (for white and rosé wines at least), it's time for fermentation. *Saccharomyces cerevisiae*, the same yeast species used for ale and bread production, is the most common yeast used to ferment grape juice as well. *Saccharomyces bayanus* is another common yeast species used for wine production and one that particularly likes fructose, which is found in roughly equal concentrations to glucose in ripe grapes and may even exceed glucose concentrations in overripe grapes. There are plenty of wild, non-*Saccharomyces* yeast strains that are commonly found on grape skins as well, so care must be taken to limit their impact on the wine. Otherwise, excess amounts of acetic acid (the acid in vinegar) or ethyl acetate (an ester common to all alcoholic fermentations that is fruity at low concentrations but reminiscent of nail-polish remover at higher concentrations) may be produced as well as some funky medicinal or barnyard characteristics that are produced from *Brettanomyces* yeast. For white wines, the style of wine being produced typically dictates the fermentation vessel used. Aromatic whites like a Sauvignon Blanc are usually fermented in a tank (typically stainless steel), whereas more full-bodied whites, like a white Burgundy, may be fermented in a barrel. White wines are generally fermented until all the sugars are depleted unless a sweeter wine is being made, in which case the fermentation will be arrested at the desired sugar concentration. Red wines are fermented on the skins and seeds and are not pressed until the desired flavors and color are achieved. They are almost exclusively fermented dry to remove all fermentable sugars (again, unless a sweet red wine is being produced). This stabilizes the wine and prevents infection from wild yeasts and bacteria, which could otherwise consume the residual sugars and produce acetic acid.

Once alcoholic fermentation is completed, many wines will then undergo a secondary malolactic fermentation. As discussed above, this is when *Oenococcus oeni*, a lactic acid bacterium, converts malic acid to lactic acid. Malic acid can be pretty pungent, with a harsh acidity, while lactic acid is less pungent, with a softer acidity that can also lend buttery, yogurt, and fruity notes to wine. This secondary fermentation is done on all red wines and with some white wines like Chardonnay or other oak-aged wines. Not only does malolactic fermentation lend different flavors and softer acidity to wines, it also stabilizes the wine by removing malic acid, which can serve as a carbon source for spoilage bacteria. Malolactic fermentation can be done sequentially after the alcoholic fermentation is completed, or it can be done by co-inoculating the *Oenococcus oeni* bacteria about 48 hours after the onset of primary alcoholic fermentation with yeast. Co-inoculation is generally the favored method, as it reduces the time required to ultimately stabilize the wine, which reduces the possibility of an infection. This also saves time and energy during production and tends to result in fresher tasting, more fruit-forward wines.

More Clarification and Maturation

After fermentation, the wine must be clarified a second time, this time to remove dead or dormant yeast and/or bacterial cells, proteins, and tannins. This helps not only aesthetically but also further stabilizes the wine by removing microbial cells and nutrients that could feed any potential spoilage organisms. Once again, clarification can be achieved through gravitational settling, often with the help of fining agents such as bentonite or isinglass (primarily collagen, made from the dried swim bladders of fish). Alternatively, filtration or centrifugation are the more expensive but also more efficient methods for clarifying wine.

Now the wine is ready for maturation. This generally occurs in oak barrels, but it may also be done in a tank with oak staves added to the tank. White wines may also be matured for a time (sometimes up to 9 months in certain styles) on the yeast lees with frequent stirring, a process known as *sur lie*, which literally translates to "on the lees," and causes the yeast cells to lyse and spill their guts into the wine. This may seem odd, since normally you want to remove these lees and not cause yeast autolysis, but this type of aging can incorporate a creamy texture and biscuit flavors to the wine. Maturation in or on oak pulls flavors from the wood and introduces

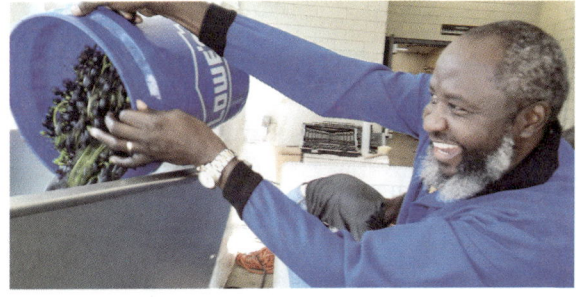

Winemaking is always a party, especially when Dr. Folarin Oguntoyinbo is involved.

Logan Isley and Megan Learn grabbing bunches of Chambourcin grapes to be destemmed.

Those same grapes in the destemmer.

What comes out of the destemmer.

The destemmed Chambourcin grapes ready for fermentation.

vanilla, chocolate, coconut, leather, and spicy notes to wines. Red wines are typically stabilized using an additional charge of sulfur dioxide prior to maturation. During maturation, they may be racked from one barrel to another multiple times to help reduce solids (tartrate salts, proteins, tannins, and microbial cells) that drop out of the wine over time. Depending on the style, the maturation period for a red wine is generally 1–2 years. Following maturation, many wines are then blended to achieve the desired flavor or tannin profile. This is most often done with red wines, but it can be done with white wines as well. The wines are racked from barrels into a tank where they can be homogenized.

Stabilization

Now are we done? Not quite yet. Now it's time for further clarification and stabilization prior to bottling. Wines must be both heat and cold stabilized. Heat stabilization prevents protein hazes from forming when the wine is warmed to room temperature. This is done by adding bentonite, which binds proteins and helps them to settle out of solution. Cold stabilization prevents tartrate crystals from forming in the wine. These are the crystals that you sometimes find in wine and that customers often confuse for small glass shards. There is nothing wrong with tartrate crystals, but at the same time, you definitely don't want your customers thinking there is glass in their bottle of wine. To cold stabilize wine, it is held at $-4°C$ ($25°$American) for 3 days, which encourages the crystals to form and drop out of solution. To test whether the wines are now cold and heat stabilized, the wine is then held at $-4°C$ for 3 days. If no crystals form, the wine is cold stable. Likewise, the wine is subsequently held at $80°C$ ($176°$American) for a few hours. If no haze forms, the wine is heat stable.

Now the wine can be filtered to fully clarify it and also to further stabilize it. If done with filter sheets with a 0.45 μm (a micron, or micrometer, is one millionth of a meter) pore size, this is considered sterile filtration as it removes just about all the remaining yeast and bacterial cells in the wine. At this point, the wine can be analyzed for pH, titratable acidity (a determination of the total acid concentration in wine), volatile acidity (the volatile acids in wine, mostly acetic acid), sulfur dioxide, alcohol, and residual sugars. Once the wine passes all quality checks, it can (finally!) be bottled. It could be released to market at this point, but some wines may go right back to the cellar for further aging in bottles prior to being released. So, yeah, for what seems like a relatively simple product, fermented grape juice, it certainly takes a lot of work and time to make a product that is suitable for consumption. There is a reason why wine is considered a luxury product with a price tag that often matches that designation.

DISTILLED SPIRITS

You may be asking, why is distillation being featured in a book about fermentation? Well, you can't have distilled spirits without fermentation! Distillation is the most effective means of concentrating alcohol in a solution. The alcohol is the product of fermentation. So, there you go.

Now that we've got that settled, let's briefly go through the process. I'm not going to get into the nitty gritty details, as that could (and has) fill up an entire book. Plus, there are many different techniques that are employed depending on the spirit being produced. For this book, I'll be focusing primarily on the historical techniques used to produce the most iconic distilled spirits in the country, which also happen to hail from the South —moonshine, Tennessee whiskey, and of course, bourbon. Besides hailing from the South, one other common factor that ties these spirits together and also distinguishes them from other spirits around the world is the corn. They all use a significant portion of corn in their grain bills. Corn, our favorite native cereal grain, is really exclusive to American distilled spirits production. No other traditional spirit from around the world uses corn as a main ingredient. And corn production, as discussed earlier, was traditionally focused in the South because of the warmer climate that didn't support other cereal grain production. And for that I say thank goodness. Otherwise, we'd probably all be drinking Scotch rather than bourbon.

In the simplest of terms, a sugar-rich solution is fermented to produce alcohol. The fermented solution, which is mostly water and roughly 5–10% alcohol with a much smaller percentage of flavor compounds, is then heated in a closed vessel so that the more volatile compounds, the alcohol and some flavor compounds, boil off as a vapor and leave the less volatile compounds behind. The vapor then flows through an apparatus attached to the vessel, where it is cooled and condensed back into a liquid. The resulting liquid, or distilled spirit, has a much higher concentration of alcohol and some of the flavor compounds. The trick is getting the alcohol concentration in the desired range and capturing the desired flavor compounds, while leaving the dangerous compounds (like methanol) and less desirable flavor compounds behind.

The technology used to make distilled spirits has been around for thousands of years and, honestly, in many applications is still used today; it has not changed much in that time. Today, the two main techniques used are batch distillation and continuous distillation. Continuous distillation is the more modern technique that uses a column still, where the distiller has much greater control over the end product. Batch distillation, by contrast, uses a pot still, and with the exception of copper and stainless steel being the preferred materials rather than clay or ceramic, a distiller from Egypt 2,000 years ago could probably figure out how to use a modern pot still after the shadow of the gnomon (look it up) has shifted a few degrees. Classic versions of moonshine, Tennessee whiskey, and bourbon were all made with pot stills and many (all, in the case of moonshine) still are.

The pot still is heated with a direct heat source (i.e., fire, the old school method) or via steam injection. The more volatile compounds boil out of the wash (more on this shortly), travel to the top of the pot still and then into the attached condenser apparatus. Classically, this would be a worm tub, which, despite the name, is not a tub full of worms. The worm is a copper coil that reduces in diameter from the top of the coil to the bottom of it. The worm sits in a wooden tub that is filled with a continuous stream of cold water from the bottom of the tub so that warm water exits from the top. As heat is exchanged from the hot vapors in the worm to the water, the vapors cool and condense and ultimately flow out of the bottom of the worm into a collection vessel.

A single batch distillation in a pot still, however, does not provide a strong enough distillate or remove enough of the undesirable compounds required for quality. So, often a doubler is added. This is just another pot still, typically smaller than the first, that captures the resulting distillate from the first pot still and redistills it to a higher alcohol content such that now different fractions can be separated and discarded or collected. The fractions are generally referred to as the heads, hearts, and tails, based on the order in which they are captured from the condenser. The heads contain a high percentage of methanol, which will cause you to go blind and then eventually kill you, so those are always discarded (I hope). The hearts are the purest fraction of ethanol and are always retained. The tails are the fraction with the least volatile congeners, some of which have positive flavor profiles and some of which have less positive flavors. As such, this is the fraction that requires the greatest mastery of blending. Much of the tails may be discarded, but at least some are partially blended with the hearts to

Always remember to check for small
students after cleaning your still.

achieve the best overall flavor for the distillate. At this point, for any whiskey, that is, the distillate is proofed (diluted to a specific alcohol percentage) and added to a barrel for aging and maturation.

The distilled spirits discussed herein all start their lives effectively as beer. Not necessarily a beer that you'd want to drink, but for all intents and purposes, it's beer. A combination of malted and unmalted grains (for the products in this book) are milled and hot water is added to the grains to make a porridge like mixture called the mash. Malted grains have been partially germinated to synthesize and activate enzymes that degrade the cell wall material and protein matrix that hold the starch granules in place in the grain endosperm. The enzymes that degrade the starch to simple sugars are also synthesized and activated, but the germination process is stopped before they have a chance to break down much of the starch. If the starch were to be degraded to simple sugars during the malting process, the grain would use the sugars as fuel to grow. This would be a loss for the distiller, as they wouldn't be able to use the sugar to feed the yeast and produce ethanol. Instead, those starch-degrading enzymes are stopped before they can go to work, and they are preserved during the kilning step in the malting process so that they can then be activated later by the distiller during the mash. Once water at the correct temperature is added, the enzymes are activated, and the starch molecules are broken down into simple sugars that yeast can metabolize into ethanol.

This is where beer production and spirits production diverge. For spirits production, the mash is cooled and yeast is added to the mixture to ferment the sugars. There is no boiling step following the mash, as there is with beer production. Rather, the product of fermentation, the wash (or beer, depending on the country in which the distillate is being produced) is fed to the still. The solids may or may not be separated from the liquid prior to distillation, just depending on the product being made and the preferred methods at the distillery.

Bourbon

To legally label a product as bourbon, there are several criteria that first need to be satisfied. It must be produced in the United States, although it does not need to be produced in Kentucky, even though the vast majority of bourbon is. The mash must consist of at least 51% corn. It can be more than that, and typically is, but it cannot be less. The distillate can be no stronger than 160 proof (80% ABV). This is a fairly low alcohol content for a distilled spirit. The reason for this is to preserve more flavor. The stronger the distillate, the more flavorless ethanol it contains. The lower proof allows for more flavorful congeners in the spirit.

Likewise, the fill strength, the proof at which the distillate is put into barrels for aging, cannot exceed 125 proof. This is somewhat dilute as far as alcohol content as well. On the one hand, a more dilute distillate at this stage occupies a greater volume, which in turn requires more barrels and more space to store the barrels. On the other hand, a more dilute distillate leads to greater flavor addition from the barrels, as most of the flavor compounds from the barrel are more soluble in water than in alcohol. You've got to love an industry that regulates quality over efficiency. There's a reason why bourbon is one of the most highly regarded whiskeys in the world (the best, in my humble opinion).

Speaking of barrels, bourbon must be aged in new, charred oak barrels. This differentiates bourbon from most other whiskeys, which allow for used barrels for the aging of their spirits. However, there is no time minimum for aging bourbon. It can be poured through a new, charred oak barrel into the bottle, which would be considered sufficient. Bourbon cannot contain any added flavoring or coloring, though, so all the color and flavor must come from the barrel, not from any type of added extract. That generally prevents the pour-through method and requires some barrel aging to

Toasting vs. Charring

Barrels may be toasted or charred prior to filling with distilled spirits. Toasting involves heating the internal surface of the wood without igniting the wood. This can be done with fire or with something like infrared radiation. It just depends on the cooper making the barrels. Regardless, lower heat is generally applied for longer periods of time, which allows the heat to penetrate deeper into the wood. Charring is when the internal surface of the barrel is ignited to produce a layer of char, or elemental carbon. The carbon layer provides little to no flavor on its own, but it does work to remove certain flavor compounds. There is typically a layer below the char layer, however, that is similar to the result of toasting.

What is the purpose of toasting and charring barrels? Well, historically charring would be done to remove any unwanted flavors carried over from the previous cargo in the barrel. Considering most foods and beverages were stored in barrels back in the day, it's no surprise that barrels would be reused and that a whiskey producer probably wouldn't want their product to taste like the salt cod that had previously been stored in the barrels. Nowadays, thankfully, it's purely to add color and flavor to aged spirits.

Cellulose and hemicellulose are the polymers (long chains of molecules) that provide structural integrity to any plant material, including the oak trees that are used to make whiskey barrels. These polymers are just long chains of sugar molecules. It takes a lot of energy to break these polymers apart, which is one of the main reasons we don't have a viable biofuels industry based on all the available biomass on the planet. Well, what has a lot of energy? Fire does, that's what! The degradation products of cellulose and hemicellulose provide flavors of butterscotch, maple, brown sugar, caramel, and nuts. Lignin is another polymer in plants that provides structural support, generally by cross-linking to the hemicellulose. It's much messier and more complex than cellulose polymers. When the lignin breaks down, it provides vanilla flavor from the compound vanillin, and spicy clove-like notes from eugenol and similar compounds. Finally, oak has compounds referred to as oak lactones that contribute coconut and celery flavors. These oak lactones are more pronounced in fresh and lightly charred barrels than in highly charred ones.

Overall, the more highly charred a barrel is, the less flavor active compounds will be imparted to the spirit. Rather, the char layer will actually remove flavor compounds, much like a charcoal filter would. This is important for removing sulfur compounds in particular. A lighter char or toast will impart more vanilla and coconut flavors, whereas a heavier char will result in a spicier character from the degradation products of the lignin. Oftentimes, barrels are toasted and charred. The toasting provides a thick layer of flavor-active compounds that are imparted to the aging spirit, while the char will provide a surface layer of carbon that imparts some flavor but also the filtering effects of elemental carbon. •

occur. To be labeled "straight bourbon," the spirit must be aged in barrels for a minimum of 2 years. And if the bourbon is less than 4 years old, the label must include a statement regarding the age.

The spirit is typically further diluted following aging. If the spirit is more dilute when it goes into the barrel, it will require less dilution prior to bottling, which showcases all those delicious congeners added from the barrel. Bourbon must go into the bottle at no less than 80 proof (40% ABV). Once all those boxes are checked, the whiskey can legally be called a bourbon (Dietsch 2016; Veach 2013).

Tennessee Whiskey

Tennessee whiskey is very similar to bourbon. The mash must consist of at least 51% corn and the spirit must be aged in new, charred oak barrels. There are two things, however, that differentiate Tennessee whiskey from bourbon. For starters, the whiskey must be produced in Tennessee. Makes sense. The other thing that separates the two whiskeys is the Lincoln County Process (see above for a more detailed discussion of the history of this process). This is the process whereby the distilled spirit, prior to diluting it to fill strength, is filtered through charcoal, typically charcoal made from sugar maple. The filtering removes many flavor-active congeners, producing a smoother, mellower tasting spirit. Following filtration, the spirit is diluted and added to new, charred oak barrels for aging. So, other than the Lincoln County Process, Tennessee whiskey is pretty much the same thing as bourbon.

Moonshine

So, I'm going to be completely honest here and state that I regard moonshine about as highly as I do hazy IPAs, which as you've probably already surmised, is not very highly. Why? Because I like to actually appreciate my foods and beverages and not feel like I'm dissolving likely important internal organs in the process. I know I know. I haven't tried your Uncle Bubba's hooch yet. Sure, there are some decent versions out there that, if distilled again and aged in oak for a couple years, would make a decent whiskey. But, in their current unrefined state, moonshine is just a bit too harsh and hot for my liking. That said, I cannot do a section on distilled spirits in the South and not cover moonshine. Especially considering that I live in the mountains of Western North Carolina, pretty much the birthplace of moonshine as we know it.

Technically speaking, there is no set of regulations, recipes, or rules that defines moonshine as a spirit. As the name references, the only real parameter that makes a spirit moonshine is the fact that it was made illegally by the shine of the moon. Sure, there are plenty of legally produced products out there nowadays that call themselves moonshine or white lightning, but technically speaking, they really aren't—other than in name—since they were made legally and were taxed according to regulations set forth by the TTB. That said, there are certain ingredients and techniques that are largely shared by most moonshiners that are also employed by these rascally legal distillers. Those are what I will cover now; but remember, if you find yourself in the middle of the woods one evening distilling something you shouldn't be, then you're a moonshiner, regardless of your grain bill or still design.

The grain bill for moonshine often resembles that for corn whiskey, which requires at least 80% corn, and the remainder is typically made up of malted barley. Regardless of the exact composition of a moonshine grain bill, it is almost always made up of at least 80% corn, hence the common reference to a corn mash. The malted barley provides the starch-degrading enzymes (to degrade the starches in the corn that is typically not malted and does not have the enzymes necessary to degrade the starch contained in its kernels). Many moonshiners will add a significant amount of corn sugar to their mash as well; corn sugar is cheap and requires no enzymes for conversion, since it's already a highly fermentable simple sugar. Many modern moonshiners will also use purchased enzymes to help with the conversion during the mash, which at least partially negates the need for malted barley. This allows the distiller more flexibility with respect to flavors when it comes to the grains used, as less emphasis is placed on the enzymatic power of the grain. Whereas corn whiskeys can be aged (although they don't have to be, and if they are, based on the grain bill and other regulations discussed above, they could then be considered bourbon) in new or used white oak barrels, moonshine is not aged.

The mash for moonshine is similar to other whiskey and beer production, where the grains are soaked in hot water to initiate and maintain the enzyme activity that converts the starches into fermentable sugars. With such a high percentage of corn in the grain bill, however, it is necessary to "cook" the corn separately from the other grains to gelatinize the starch in the corn. Starch is found in ordered, crystalline granules in grains. For the enzymes to be able to access the starches and break them down to

247

sugars, the starch granules must first be opened up. This is the gelatini-zation process. Corn happens to have a higher gelatinization temperature than barley, wheat, rye, and oats, the other common grains used in distill-ing. The typical mash temperature for maximizing enzyme activity is 65°C (149°American). This is plenty warm enough to gelatinize starches in the other commonly used grains, but not in corn, which requires temperatures of at least 70°C (158°American) and possibly up to 80°C (176°American). So, the corn is typically "cooked" separately at those elevated temperatures or even boiled for about 30 minutes. The corn must then be cooled prior to adding the other grains so that the enzymes in the other grains are not inactivated. You know what they say, "Inactive enzymes don't convert," and "A denatured enzyme can never be activated again." They don't say that, you say? Well, they should, because it's true.

Once the enzymes have done their thing and the starches have been converted to fermentable sugars, it's time to ferment. Unlike with beer and some other distilled spirit fermentations, in which the liquid is separated from the grains prior to fermentation, for moonshine production, the mash is cooled to fermentation temperatures and the yeast is pitched directly into the mash, solids and all. The most commonly used yeast is distiller's active dry yeast, which is a fast-fermenting, alcohol tolerant *Saccharomy-ces cerevisiae* strain developed specifically for fermenting distilled spirit grain mash fermentations. There are plenty of other options out there as well, however, depending on your specific needs and preferences. The yeast is typically complemented by the addition of yeast nutrients to provide any additional nitrogen required (although this shouldn't be necessary with an all-grain mash) as well as important vitamins and minerals. Fermentation temperatures are generally warmer than would be used with most beers, in the 24–27°C (75–81°American) range, as the emphasis is on an accelerated rate of fermentation (when the feds are hunting you, it's difficult to be lei-surely about the process) and alcohol production, not the resulting flavor profile of the ferment, as this can be accounted for during distillation. At these temperatures and with the proper yeast and nutrients, fermentation should be completed within a few days.

At this point, the liquids can be separated from the solids (and the re-sulting solution would be referred to as the wash), or not. That's right, many distillers run the whole shebang, grains and all, through their still. Other-wise, the wash is introduced to the still without the grains. It's definitely less messy to distill the wash, but it depends on your preferred technique and the capability of your still and overall setup. Moonshine is distilled exclusively on pot stills. The first pass through the still, in which no fractions are removed and the entire distillate is collected until a very low ABV is measured (down to 10% ABV or less, oftentimes), is referred to as the low wines. These will generally have an overall ABV of about 60%. The low wines are then redistilled and the different fractions, the heads, hearts, and tails, are collected separately. The hearts are always kept, as they are primarily ethanol. Most of the heads are discarded, as they contain methanol and other harmful compounds, although some can be blended in for their fruity flavor. The tails are lower in alcohol and contain the fusel alcohols (butanol, propanol, etc.), fatty acids, fatty acid esters, and other less volatile compounds. Most of these are also discarded, as they can have quite an unsavory character, but blending in some small amount of the tails can add character to the distillate. If your setup has a doubler, then the doubler is doing the redistillation of the low wines for you and the whole process can be knocked out in one run. The resulting moonshine, following the redistillation of the low wines and blending of the desired fractions, will typically have a high proof between 100 (50% ABV) and 180 (90% ABV) and is now ready to drink (or remove paint, depending on your preference)!

Chapter 18

PRESENT

The current fermented beverage landscape is rich all over the country, but there are particularly notable hotspots in the South for certain beverages. Asheville, North Carolina, is one of the craft beer capitals of the country. Kentucky, Virginia, North Carolina, and even Florida are some of the top wine-producing states nationwide. I will share an insider's view of many of the top breweries and wineries in the South. And then there are, of course, the distilleries that turn fermented beverages into masterful distilled spirits. Namely, of course, there are Kentucky and Tennessee for bourbon and Tennessee whiskey, respectively. Heck, many of my former students are the head brewers, winemakers, and distillers at these businesses, so this chapter will provide an excuse to pay them some visits.

But the fermented-beverage landscape stretches so far beyond these industries nowadays to more alcoholic beverages like cider and mead as well as to non-alcoholic beverages like kombucha, jun, and vinegars. I have to admit that I missed the mark about 10 years ago with regard to cider and mead. I was sure that mead was going to be the next big thing as gluten-related illnesses became more readily diagnosed. Alas, cider was the beverage of choice that stepped in to fill the gluten intolerant void.

Kombucha and jun have become very popular as regularly consumed beverages, and these as well as vinegar, which may not be a beverage that one would regularly consume (other than for some potential health benefits), have also found a place in many chefs' culinary toolboxes. The acid is an obvious reason, but the depth of flavors produced through fermentation provides a veritable cornucopia of options for sprucing up a dish with a little dash. Nonalcoholic fermented beverage sales are growing nationwide and continue to grow in popularity in the South as well.

THE CURRENT BEERSCAPE OF THE SOUTH

Beer is big business in this country. The South may have been a little late to the craft beer party, but boy have we made up for lost ground since we entered the fray. There are currently just under 10,000 breweries in this country, an all-time high. The previous high, before the last decade, was in the late nineteenth century, when there were just under 5,000 breweries. Granted there are a lot more people in this country now than back then, but still, that's pretty solid growth, especially considering we were down to fewer than 100 breweries in the entire country from the late 1970s into the early 1980s. The last decade alone has seen the number of breweries almost quadruple. The vast majority of this growth has been in craft breweries. Even so, craft beer accounts for only about 13% of the overall volume of beer produced (economically speaking, craft accounts for almost 25% of the beer market, however). As far as the South is concerned, Texas, Florida, and North Carolina lead the way in overall number of breweries as well as in total economic impact, a figure that includes all revenues from brewery products as they move through the three-tier system (breweries, wholesalers, and retailers) as well as from non-beer products, such as food and merchandise, sold in the breweries. To give you an idea of the economic impacts we're talking about, beer accounts for just under $5 billion in Texas and just over $2 billion in North Carolina. That's a lot of tax revenue for the states and the feds both, which is why beer has a pretty strong lobby in Congress these days. I'm particularly proud of my home state. While we may rank third overall in the South, that's third compared to Texas and Florida, two giant states. If you break it down to breweries per capita, North Carolina leads the way in the South. If you go to Asheville, or even to Charlotte or Raleigh these days, it's hard to walk a few minutes in any direction and not end up in a brewery. Yup, it's a good time to be alive.

It's no surprise, either, that North Carolina is a leader in this category. A lot of that success can be traced back to the mountains of Western North Carolina. Asheville was a huge player in the initial phases of growth in the craft beer scene. And it all started when Oscar Wong opened Highland Brewing back in 1994, way before craft beer was even on anyone's radar. Oscar, the venerable godfather of the Asheville brewing scene, is not necessarily who you might picture when you think of the guy who kicked off one of the biggest and best brewing scenes in the country. Oscar was born and raised in Jamaica and then emigrated to the United States to attend Notre Dame where he earned a master's degree in civil and structural engineering. After a long and successful career as an engineer and founder of a nuclear waste processing company in Charlotte, he retired to Asheville in 1992 to settle down, live the good life, and—what else—brew beer.

The avocation grew into a vocation when he rented the basement under Barley's Taproom in downtown Asheville to open Highland Brewing in 1994. The brewery, like many others back in the day, was constructed mainly from retrofitted dairy equipment. Nonetheless, the space was 12,000 square feet, which would make many current brewery owners seriously jealous, and they were able to crank out 6,500 barrels of beer (a barrel, or bbl, is 31 US gallons), which, again, would be the envy of many current craft brewery owners. What started as a way to ensure that Oscar would always have a place to go for a tasty beverage quickly grew into an iconic brewery in an already popular tourist destination that was on the brink of exploding onto the craft beer scene. In 2006, Highland Brewing moved to its current location in East Asheville, a 40-acre property that was the former home of Blue Ridge Motion Picture Studios. Highland Brewing has continued to expand and now has a 50 bbl brewhouse in a 180,000 square foot building with a taproom, packaging facility, event center, and rooftop bar.

In 2015, Oscar finally decided it was time to retire (for real this time . . . maybe) and Leah Wong, his daughter, took over as president and CEO. Recently, Oscar was invited to attend the Boy Scouts 2023 Citizen of the Year dinner where he was awarded the honor by the Asheville area Boy Scouts of America. What he wasn't expecting was the award bestowed upon him by the governor of North Carolina, the Order of the Long Leaf Pine, the highest civilian honor awarded by the governor (Kennell 2023). I could not agree more with the governor's choice for this prestigious award.

With nearly 60 breweries, many people will tell you Asheville has the most breweries per capita of any city in the country. It doesn't, but they most certainly have a lot of breweries. And with so much competition, most of the breweries are pretty darn good, too. Asheville was actually voted Beer City USA for 4 years in a row, an award determined by an online poll conducted by the Brewers Association. This may have been over a decade ago, and some city up in Michigan may have won it since, but nobody down here cares. Asheville is, and always will be, Beer City to those of us who call the South home.

It's such a beer hot spot, in fact, that White Labs, one of the largest yeast labs in the world, and the first to produce a pitchable liquid yeast culture for brewing, opened up an East Coast facility in Asheville in 2016. Sure, the facility was opened to improve East Coast distribution of their products in general, and Asheville made sense because it's rather centrally located on the East Coast with easy access to a major highway, but the fact that there was a major brewing scene in Asheville certainly didn't hurt. White Labs in Asheville now also has its own brewery and restaurant too, which cranks out amazing food, including lots of fermented goodies like kimchi fries, made with fries that have been lactic acid fermented and topped with house-fermented kimchi. If you haven't been to Asheville yet, you need to go. There are so many great breweries and restaurants, it's hard to keep track of them all now. My only advice is to bring your appetite!

NEAREST GREEN FOUNDATION AND DISTILLERY

While traveling in Singapore, Fawn Weaver, entrepreneur and best-selling author, read a *New York Times* article about Nearest Green and how he was the secret ingredient behind Jack Daniel's success. This led her to Lynchburg, Tennessee, followed by thousands of hours spent researching Uncle Nearest and his legacy with Tennessee whiskey. Weaver ended up purchasing the 300-acre farm in Lynchburg (where Nearest taught young Jack Daniel how to make the best whiskey around), and she started the Nearest Green Distillery in Shelbyville, Tennessee. She serves as the CEO of Uncle Nearest Premium Whiskey, the fastest-growing American whiskey brand in US history, the best-selling African American–founded spirit brand in history, and the most award-winning whiskey of 2019, 2020, and 2021. She also employs an all-female executive team and Master Blender Victoria Eady Butler, a fifth-generation Nearest Green descendant. Oh yeah, and Weaver also started the Nearest Green Foundation, a 501(c)(3) nonprofit organization that provides full college scholarships to any descendant of Nearest Green (Barthole 2022; Fluker 2022; Pharms 2022).

The story is amazing. But how's the whiskey, you ask. Well, I can say after many hours of in-depth research myself that it is, hands-down, one of the better whiskeys I've tried in a long time. I'm not usually a fan of Tennessee whiskey, as I find it overly sweet. Not Uncle Nearest. Ironically, my brother in New York is the one who turned me on to it. I've since tried the 1856, the 1884, and the rye, which are all excellent. I suggest you run out immediately and pick up a few bottles for yourself (but make sure to leave some for me).

Chapter 19

FUTURE

—

YEAST HYBRIDIZATION

You've probably already mentally blocked from your memories that torrid yeast love affair excerpt presented earlier in chapter 17 (either that, or it's indelibly imprinted on your memory and you will forever curse me for subjecting you to that), so you probably don't remember when Johann the yeast cell asked Inez to shmoo with him. Or you just figured it was some weird term of endearment that I like to use. Regardless, "shmoo" is an actual scientific term from microbiology. It refers to a "mating projection" (yeasts are notoriously repressed about such appendages) that grows on the haploid spores (those with a single set of chromosomes) of yeast cells that form during the sexual reproduction phase. The shmoos come together and fuse to form a diploid (two sets of chromosomes, one from each spore), a process called shmooing (I swear I'm not making this up). *Et voilà*, a new hybrid yeast cell is born, much like the daughter cell made by Johann and Inez earlier. Now, for those of you who are of an age, you may remember the Al Capp cartoon *The Shmoo*, which first appeared in the *Li'l Abner* comic (which, if you're not aware, was one of several denigrating cartoon portrayals of the South that were popular back then). Shmoos are friendly creatures that multiplied (asexually, mind you) faster than rabbits and that required no food or water, only air to survive and thrive. They are delicious and absolutely love to be eaten. So much so that if a human looked at one hungrily, it would happily jump onto any cooking apparatus available. If thrown on the grill, they taste like steak; in a frying pan, they taste like chicken; roasted, yup, pork; and baked, well, catfish, of course. If you eat them raw, they taste like oysters. I would like to remind you that I'm not making any of this up. OK, so hopefully you get that this was a very weird cartoon character at this point. But, now to my point for this whole bizarre tale. Their appearance was simplistic, but certainly unique. They looked like white, plump bowling pins with short little legs, no arms, and just big eyes, a mouth, and whiskers (Berger 1970). They looked not unlike a cartoon version of the sexual haploid yeast spore with a shmoo. And there's a good reason for that. The microbiology term "shmoo" was actually taken from the cartoon character because of its similarity in appearance.

Yeast hybridization is the somewhat rare sexual reproduction between yeast cells that you were unfortunate enough to witnesses with Johann and Inez. In nature, yeasts almost invariably reproduce asexually by budding, such that the daughter cells have an identical genome as the parent cell. However, when stressed from harsh environmental conditions, such as lacking necessary nutrients, yeasts can sexually reproduce to increase the genetic diversity in the population and to increase their chances of survival. Well, such conditions are easy enough to replicate in the lab, and yeast cells can be encouraged to sexually hybridize. The reason for the yeast doing so is the same as in a natural setting, but the reason for scientists to encourage the process is to create a genetically novel yeast strain with certain desirable characteristics, some of which are inherited from each parent. It's certainly no surefire thing that the hybrid yeast will have the intended characteristics, so the process is typically done numerous times before an acceptable hybrid is developed.

Even after a hybrid yeast with the desired traits is produced, it must go through a litany of tests to ensure that it will still ferment adequately, flocculate (settle out of solution) properly, and demonstrate that it's genetically stable. The benefit of this method over, say, genetically modifying yeast by swapping genes with other microbes using genetic engineering techniques is that, for one, the yeast certainly enjoy it more. At least I assume they do. Seriously, though, even though promoting the sexual hybridization of yeast is also genetically modifying the yeast to produce new cells with desirable characteristics, it is considered to be done "naturally" and, therefore, still considered GRAS (generally recognized as safe) and not subject to any additional regulatory clearances. Plus, there is no public stigma associated with this process like there is with GMOs. The process is already being used with success to produce many different yeast strains with interesting and desirable characteristics.

The reason I'm suggesting this is the future is because there is still so much untapped potential to unlock with yeasts. If you think about the other ingredients in beer—water, malt, hops—there is really only so much that can be done with them to fundamentally change and advance beer production. There could certainly be advances in the technology of beer production (equipment, processes, etc.), but other than stainless steel and pumps, not much has changed in the last couple thousand years. Plus, I'm a scientist, not an engineer, so I'm sticking to my scientific guns here. Yeasts are responsible for the fermentation process and so much of the flavor profile and overall characteristics of the final product. And we're just beginning to understand how to truly tap their potential. That said, there is only so much change that can be made to yeast using sexual hybridization. The other option with even greater potential for impact is genetic modification using genetic engineering.

The reason I'm not automatically hanging my hat on genetic modification through genetic engineering is primarily because of consumer acceptance, or lack thereof. Many people hear GMO and freak out. And many of those people are also the most avid craft beer consumers. That's not to say that some yeasts have not already been genetically engineered. There are yeasts that have been genetically modified to produce lactic acid along with ethanol, so that the brewer can easily produce a sour beer without using kettle souring or a long, mixed-culture fermentation. Other yeasts have been genetically modified to release thiols (sulfur containing compounds that have a strong, tropical fruit character—think Sauvignon Blanc) into beer from their bound precursor molecules. Still more yeasts have been modified to produce the terpene molecules found in the essential oils of hops (that provide the hop aroma and flavor to beer), so that a beer could potentially be made with no hops. A yeast that can produce cannabinoids? Yup, those have been made in the lab as well. Yeasts, like other microbes, can effectively be turned into tiny little factories to produce pretty much any compound desired with the right genetic tweaking. I don't have a problem using these organisms to produce beer, but I also don't have to worry about selling my beer to consumers. So, regardless of which technique ultimately rules the day (or whether the burden will be shared), I feel strongly that genetically modifying yeast, either through more "natural" techniques or through genetic engineering, will have a huge impact on the brewing industry in the coming decade or two.

AMERICAN SINGLE MALT WHISKY

You may have heard of American single malt whisky (the official name leaves out the "e" in the spelling as an acknowledgment of the more universal spelling—see sidebar 17.1 for an explanation of the different spellings of the word) and hopefully have already tried it, but I put it in the "Future" chapter because, as of this writing, it is a category that has yet to be officially recognized by the Alcohol and Tobacco Tax and Trade Bureau (TTB). If you're not familiar, the TTB, part of the US Department of Treasury, regulates and collects taxes on trade and imports of alcohol (and tobacco and firearms) in the United States, so they get to say when a category of whiskey is officially recognized. It is coming, though (and may already happen by the time of publication). The distillers are proposing to establish a standard of identity for American single malt whisky, which would require the product to be distilled entirely at one US distillery and also mashed, distilled, and aged in the United States. Additionally, the mash bill must consist of 100% malted barley, the spirit can be distilled to no more than 160 proof (80% ABV), matured in oak casks with a capacity not exceeding 700 L, and bottled at no less than 80 proof (40% ABV).

These standards may sound very similar to those for Scotch, particularly single malt Scotch, which must also come from a single distillery and use 100% malted barley. OK, so the difference must be the peat, right? Not so fast. First of all, not all Scotches are peated. There is no legal requirement that it is, and there are plenty of unpeated Scotches out there. Second, there are several American single malt whiskies that are peated, although they are less common than their Scottish counterparts.

One difference is in the ages of the whiskies. Scotch has a minimum aging requirement of 3 years, whereas American single malt does not (or will not once it's officially recognized by the TTB). You will typically see the ages on Scotch labels too. This is done mainly for marketing reasons, as people generally equate longer aging times for whisky with higher quality. And while the more time a whisky spends in a barrel and the more evaporation that occurs, the more valuable the remaining spirit becomes; but it does not necessarily mean that the whisky is any better from a quality standpoint. I'll probably be banned from any online whisky forums for all eternity for writing this, but I actually prefer the younger Pappy Van Winkle bourbon to the more aged options. I've been fortunate enough to try the 15-, 20-, and 23-year-old Pappy's, and the 15 was my favorite. Granted, 15 years is still pretty aged when it comes to bourbon, but it still retained some of the original wheated bourbon character, whereas the older offerings were so oaky and smooth that, in my opinion, the bourbon character was lost. They were more one note compared to the more complex 15-year-old version. Not that I can afford any of them, but if I could, I would stick to the younger one for drinking and have the more aged versions in my liquor cabinet only to show off to people (which is what people do with Pappy's anyway).

Part of the reason why Scotch requires a minimum aging time is because whisky "ages" much more slowly in Scotland than it does in the states. The drastic temperature swings we get in the South between the seasons produces similarly drastic "breathing" of the whisky barrels, as the wood pores grow and shrink depending on the temperature, which ages the whisky rapidly, versus Scotland, where it's pretty much 10°C (50°American) year-round, which creates an environment in which the whisky ages much more slowly. Other than that, the main difference between Scotch and American single malt whisky is that one is produced in Scotland and the other in America. In fact, I'm working on converting my father from a Scotch drinker to an American single malt drinker. Although he may be a lost cause, as his whisky of choice is Johnny Walker Black. Oh well, I guess it could be worse. He could prefer the Red.

Virginia Distillery Company

One American single malt distillery that stands out in my opinion is the Virginia Distillery Company, a woman-owned, women-made company, in Lovingston, Virginia, and not only because they just endowed the Angela H. Moore Women in Distilling scholarship (named after the owner) to support students in our program and encourage more women to get into a currently male-dominated industry. The company also happens to be the largest independently owned (and woman-owned!) producer of American single malt whisky in the country, and by the way, they make really good whisky. I was fortunate enough to tour their beautiful facilities recently. They imported the majority of their equipment, including their grain mill and pot stills, from Scotland, and they use a process identical to Scotch production so that if you ignored the lack of funny accents and overcast, chilly weather, you might think you were in Scotland.

And it works. I was able to taste the three of their Courage and Conviction line of whiskies straight out of the barrel, aged in three different types of barrels. The Bourbon Cask whisky, which was aging in barrels sourced from Kentucky that previously held bourbon, was smooth and rich, with notes of vanilla and butterscotch and a hint of smoke. It was what you might imagine if a nice, smooth bourbon and a very lightly peated Scotch had a delicious love child. Their Sherry Cask whisky is aged in former Sherry casks that once held Fino, Oloroso, or Pedro Ximenez sherries and definitely picked up a lot of the sherry characteristics, with those classic nutty and pit fruit flavors. My personal favorite was the Cuvée Cask whisky, which is aged in barrels that once held European red wines. However, before being filled with whisky, the barrels are shaved and re-toasted and re-charred. It was a deliciously smooth whisky with ripe berry and pecan notes shining through.

They are even producing a whisky specifically to support the Women in Distilling scholarship, the Scholar's Craft. The whisky was aged in former bourbon barrels and then finished in barrels that previously held ethically sourced, small-batch coffee. Not only is the whisky supporting a great cause, it is also objectively delicious. Imagine an unpeated or lightly peated Scotch that finishes with a chocolatey coffee flavor. The coffee is an acknowledgment to the role that coffee plays in the lives of the students the scholarship is supporting. I won't tell them that most undergraduate students that I know generally prefer those awful, sugar-bomb energy drinks to quality coffee. In the distillery's defense, though, a whisky aged in Red Bull barrels probably wouldn't have quite the same appeal. Anyway, check it out. You can sip fine whisky while knowing that you're also supporting a good cause.

Chapter 20

RECIPES

―

....

KOMBUCHA

Prep time: 15 minutes | Fermentation time: 1–2 weeks | Yield: 4 L

Whether or not you're a fan of the strange wheybucha concoction introduced in chapter 15 of this book, at least you're already familiar with the
process. Here, we're essentially going to use the same process, but this
time we'll use it to make a much more traditional kombucha. This is a
health tonic that has been around for millennia, likely originating in China
and spreading west to Russia from there. It started to gain popularity in the
United States in the early 1990s, but really took off in the last decade or so.
It's one of the easiest things to ferment at home, which is great because
commercial kombucha can be quite expensive. So, here we go.

Ingredients

- 4 L filtered water
- 20 g (6 tea bags) black or green tea
- 300 g sugar
- Kombucha starter SCOBY or about 60 mL of fresh, live culture
  kombucha

• Optional spices or fruit: try adding some fresh ginger (about a finger's worth) and a hot pepper—my personal favorite, other spices of your choosing, or some fruit (mango is yummy!) to jazz it up. I would start with the basic ingredients first, though, and once you get the process down, then you can start adding in other ingredients for fun.

Instructions

Pour the water into a pot and heat it up until it just starts to boil.

Turn the heat off and stir the sugar in until it is fully dissolved.

Add the tea bags to the sugar water and steep them for about 15 minutes.

Remove the tea bags and pour the solution into a glass jar or other fermentation container.

Let the solution cool to below 37°C (100°F) if not room temperature.

Add the starter SCOBY or a dose of fresh, live culture kombucha to the solution.

Cover the jar with a towel or coffee filter and secure with a rubber band.

Let the kombucha ferment at room temperature for at least 1 week, or up to about 2 weeks. Check the pH after the second day to ensure that it is below 4.6. It should drop below 4 to about 3.5 or so by the time it is finished. After 1 week, taste it to see if it is acidic enough for you and ready to drink. The longer you let it go, the more acidic it will become, so if you want a kombucha with a little more sweetness, then you probably want to stop it after 1 week or so. Also, the warmer it is, the faster it will ferment, so keep that in mind as the internal temperature of your home changes seasonally.

Once it's done to your liking, you can filter it to remove the SCOBY and any sediment and/or spices.

Cover the filtered kombucha with a lid and put it in the refrigerator. It should last for a couple of weeks in the refrigerator.

Reserve the SCOBY in a little of the kombucha you just made and store it in the refrigerator. It will last for 1 month or so without additional feeding. You can use it to make your next batch.

Fermentation

We've already discussed the microbes that are involved in kombucha fermentation and the basic process, but we haven't yet discussed the potential health benefits. The key word in that last sentence being "potential," as most studies regarding the health benefits of kombucha have been in vitro or done with animal models. There are few in vivo and even fewer (if any) clinical studies that have been done yet, so it is difficult to make any health claims with certainty. However, the evidence is pretty strong that kombucha is a very healthful drink, with at least one very important caveat. First of all, the tea itself provides lots of phenolic compounds (the tannins from the tea) that are potent antioxidants. The fermentation process produces plenty of other healthful compounds as well, including B-complex vitamins and organic acids such as acetic acid, lactic acid, gluconic acid, glucuronic acid, and ascorbic acid. Apple cider vinegar, which is about 5% acetic acid, has been found to decrease total serum cholesterol as well as blood sugar levels (Hadi et al. 2021). Glucuronic acid is a potent detoxifier that can bind toxins in the liver, allowing them to be excreted and removed (Kitwetcharoen et al. 2023). Overall, kombucha may provide antioxidant, antimicrobial, antitumoral, anti-inflammatory, antihypertensive, hepatoprotective (prevents liver damage), hypocholesterolemic, antidiabetic, and probiotic benefits (Diez-Ozaeta and Juaristi Astiazaran 2022). That said, there is that one caveat. The Centers for Disease Control and Prevention recommends that humans consume no more than about 4 ounces of kombucha per day. This is to prevent acidosis, a condition that occurs when your blood becomes too acidic. Kombucha is a very acidic drink, similar to vinegar. You probably wouldn't want to consume more than 4 ounces of vinegar a day either, so stick to the same recommendation for kombucha. It is likely one of the healthiest beverages you can consume, but in moderation.

....

WATERMELON TEPACHE

Prep time: 30 minutes | Fermentation time: 3–7 days | Yield: 1 L

Ah, tepache. One of the most refreshing fermented beverages you can possibly enjoy. It originates from Mexico and is typically sold by street vendors just about everywhere that has street vendors. It's a slightly alcoholic fermented beverage made from the core and outer skin of a pineapple. That's one of the reasons I love it so much. It's using what would normally be a waste product and turning it into a delicious, nutritious fermented beverage. Depending on where you are in Mexico, though, different flavorings, fruits, or spices will be added that reflect the flavors of that region. Well, since this is a book about southern fermentations, let's add some watermelon in there, the quintessential southern US fruit. Plus, it's delicious, so here we go.

Ingredients

- The skin and core of 1 pineapple
- The rind of 1 watermelon (if you don't use them for making pickles) and some of the fruit, too, for flavor and sugar (however much you want to add—this is a street vendor food made from waste products, so no need to get too technical, just throw in that leftover watermelon that inevitably ends up sitting in the fridge after the initial foray)
- 200 g of piloncillo or brown sugar
- 1 cinnamon stick
- Other optional spices: ginger, cloves, hot peppers, or anything else you've got a hankering for
- Filtered water

Instructions

Remove the skin and core of the pineapple, as you would normally do when cutting up a pineapple. The top can be composted, or you can use it to grow a pineapple plant. Use the fruit to make your pineapple habanero fermented hot sauce (see the recipe in chapter 5).

Cut the skin and core into 2.5–3 cm pieces.

Find that leftover watermelon in your refrigerator and cut the fruit into similarly sized pieces. The rind and skin can also be used but won't add much flavor to the process.

Put the sugar and spices into a 2 L (or similarly sized) jar and add enough filtered water to dissolve the sugar after stirring.

Add the fruit to the jar and top it up with enough filtered water.

Place a towel or coffee filter over the top of the jar and secure it with a rubber band.

Allow the tepache to ferment at room temperature for about 3 days (no additional inoculum needed, as the microbes on the skin of the pineapple will do the trick). You should see some frothy bubbles at the top of the solution. Check the pH to ensure that it has dropped below 4.6. If it has, taste the tepache to determine whether it has the right balance of sweet and sour for you. If it's too sweet, let it go another day and check it again. If you let it go too long, though, it may get overly sour. Plus, you really want to catch it a little before it's fully done, so that you can use the continued fermentation activity to carbonate the beverage.

Once it's done to your liking (or a little before), filter the tepache into swing-top bottles or something similar that can handle some pressure from carbonation.

Close the swing tops and leave the tepache out at room temperature for another day. The carbon dioxide from fermentation will naturally carbonate the beverage.

Place the tepache in the refrigerator and fully chill it before opening.

When you're ready to enjoy it, open it carefully over the sink and without sticking your face directly over it, as it may gush a bit if over carbonated.

Drink it straight or add a splash of your favorite distilled spirit and enjoy.

Fermentation

Very few scientific studies have been done on tepache, so it's difficult to de-scribe the particular microbes involved with any certainty. Plus, since no in-oculum is used in the traditional production process, the microbes involved will differ depending on the location and the particular pineapples (and/ or other fermentable material) used. There are actually at least a couple of commercial tepache producers that are distributing to the United States now, and one of the products distributed is alcoholic. No mention is made of whether the producers are using an inoculum, although I'm sure the al-coholic version is using *Saccharomyces cerevisiae* to achieve an ABV of 7%. One study out of Poland, of all places, investigated the microbes involved in tepache fermentation and found the bacteria *Lactobacillus pentosus*, *Lacto-bacillus paracasei*, *Lactobacillus plantarum*, and *Lactobacillus lactis* as well as a *Saccharomyces* yeast (Ligenza et al. 2021). These microbes not only support healthy gut microflora, they also help support the immune system and regulate digestive health. The *Saccharomyces* yeast will produce a little bit of alcohol, so most tepache is slightly alcoholic, but it rarely gets above

2% ABV naturally. Regardless, te-pache is delicious and refreshing, and if you haven't had it yet and can't make it to Mexico anytime soon, make some for yourself and imagine yourself luxuriating on a warm, sunny beach as you sip your fermented, mildly alcoholic beverage.

A jar of tepache as it ferments.

....

CAROLINA COMMON ALE MADE WITH NORTH CAROLINA SWEET POTATOES

Before I get into this particular beer, I want to quickly discuss the manner in which I'm presenting the beer recipes that I'm including. Since people trying these recipes will be using different sized systems with different efficiencies, I'll be more general regarding the amounts of materials used. Malts and other carbohydrate sources will be provided as percentages with a target original gravity for the batch. The hop additions will give the time for each addition and the total International Bitterness Units (IBUs) added to the batch. The yeasts listed all come from White Labs (www.whitelabs.com), which is always a quality source with a huge yeast and bacteria library, but Omega Yeast (https://omegayeast.com) has some great strains as well and is always doing a ton of research and development into new and exciting yeast strains.

The water profiles described will be pretty generic, too. Water chemistry is a very complex subject that I typically spend multiple lectures on in my Principles of Brewing Science course. Rather than getting into that level of detail, I will provide the target calcium ion concentration (arguably the most important ion for brewing) as well as the ratio of chloride and sulfate ions, which plays a large role in the taste and mouthfeel of the beer. When the sulfate to chloride ratio is 2:1, you get a drier mouthfeel and mineral bitterness that accentuates hop bitterness (think West Coast IPAs). When that ratio is flipped (twice as much chloride as sulfate), you get a fuller mouthfeel and a sweeter taste (think hazy IPAs). When the ratio between the 2 is balanced, you get a more balanced mouthfeel with no real sweetness or mineral bitterness to speak of. Finally, for darker beers, the water should have adequate residual alkalinity to buffer the addition of the darker malts, which are acidic. Residual alkalinity results from the amount of bicarbonate in the water that is available to react with acids. Ultimately, you want the pH of your mash in the 5.1–5.4 range (at mash temperatures; the pH will be higher at room temperature). If it's too low, you need more residual alkalinity in your water. If it's too high, then you need less. To raise residual alkalinity, use baking soda (sodium bicarbonate), not calcium carbonate (chalk), which really doesn't like to dissolve under typical brewing conditions. Just be aware that you are also adding sodium to your water. There's nothing wrong with that, and you have to add quite a bit before you can detect it. Even when you can, it adds a little mouthfeel and accentuates flavor (just like with food). To lower residual alkalinity, adding calcium (in the form of calcium chloride or calcium sulfate) will do the trick. But, keep in mind that you are also adding in chloride or sulfate and should follow my guidance above. I told you it was complicated.

Many brewers try to mimic the water profiles of the places where certain styles originated. This can be very challenging, however, especially if you already have a lot of mineral content in your water. Rather, stick to my few guidelines above and you can save yourself time and headaches trying to figure out the right mineral additions. Most states provide free or very reasonable full water analyses, typically through the state agronomy lab. Talk to your local extension agent and they can tell you how to send in a water sample. Most brewing software will calculate your water chemistry for you based on your existing water profile and any salt additions that you make.

Now that we've got that out of the way, let's get to the Carolina Common. This is one of the favorite beers I've made. It's based on the historical Kentucky Common style, which uses a good bit of corn in the mash. Since sweet potatoes are our thing in North Carolina, we replaced the corn with some North Carolina sweet potatoes. The result is malty, but still light and delicious, almost reminiscent of an English dark mild. And I love the name we gave it when we did it as a collaboration with Booneshine Brewing Company—I Drink Therefore I Yam. I should also mention that this beer won a gold medal at the US Open College Beer Championship. Enjoy!

Ingredients

- 60% 2-Row Pale Ale malt
- 20% North Carolina sweet potatoes
- 9% Dark Munich malt
- 5% Rye malt
- 5% Crystal Rye malt
- 1% Midnight Wheat malt
- Magnum hops—60 minute addition for 16 IBUs
- Saaz hops—whirlpool addition for 4 IBUs
- Fining agent and yeast nutrient optional but recommended
- WLP002 English Ale Yeast

The sweet potatoes on the brew deck at Booneshine Brewing Company, waiting to be added to the mash.

Thomas Prebeck stirring the sweet potatoes into the mash.

Instructions

The water profile should be pretty balanced for this one, ensuring that the calcium concentration is at least 50 ppm, with roughly equal sulfate and chloride concentrations and adequate residual alkalinity to buffer the darker malts.

Cut the sweet potatoes into quarters or smaller and roast them until they're done. Cool them to room temperature.

Add the sweet potatoes with the grains to the mash and use a single infusion mash at 66–67°C (151–153°F) for 60 minutes or until the starches are adequately converted.

You're shooting for an original gravity of 1.047 (11.75°P), so ensure that your kettle gravity is in the right range based on your evaporation rate.

Boil for 1 hour, adding the hops at the specified times and the fining agent and yeast nutrient with 15 minutes left in the boil.

Whirlpool for 15 minutes and let stand for an additional 15 minutes.

Chill the beer to fermentation temperature and aerate to 12 ppm oxygen.

Ferment between 18–20°C (65–68°F).

Fermentation

The WLP002 is a classic English yeast strain that originally came from Fuller's Brewery. For a classic English strain, it is a pretty clean fermenter that produces some fruity esters; but it is still a great choice for styles requiring a more muted contribution from the yeast. That said, it is typically English, as it likes to flocculate fairly early and rather completely. So, you'll end up with a nice, clear beer, but you may also have some residual diacetyl (the buttery off-flavor that is common in English styles) that the yeast didn't mop up before flocculating. And you may end up with a little residual sweetness, also resulting from early flocculation. This accentuates the malt character of a beer, which is perfect for this particular beer.

....

CHOCOLATE SALTY BALLS

This is a beer that we've been making for years. Interestingly, it all started with cookies that my wife made for her holiday cookie exchange. They were chocolate salted rye cookies that were absolutely delicious. It always makes me sad when she makes such amazing cookies for the cookie exchange because I know she's going to give away 90% of them and bring back the standard assortment of boring holiday cookies. I mean, how can you possibly exchange chocolate salted rye cookies for star-shaped sugar cookies with sprinkles and call that fair? Anyway, I digress. The cookies were so good, I knew that I needed to make them into a beer. And naming them after Chef's Chocolate Salty Balls (it's a *South Park* reference in case you're wondering) just seemed so perfect.

Ingredients

- 50% 2-Row Pale Ale malt
- 25% Rye malt
- 10% flaked oats
- 5% Caramel 60L malt
- 5% Crystal Rye malt
- 2.5% Blackprinz malt
- 2.5% Chocolate malt
- Cacao nibs (6 g per L) at 5 minutes left in the boil
- Magnum hops—60 minute addition for 30 IBUs
- Fuggle hops—whirlpool addition for 5 IBUs
- Fining agent and yeast nutrient optional but recommended
- WLP001 California Ale Yeast®

Instructions

The water for this beer should be salty. I mean, it's Chocolate Salty Balls, after all. Aim for about 100 ppm calcium; 250 ppm sodium, chloride, and bicarbonate; with 125 ppm sulfate. So, not only should the water be a little salty, but notice that the chloride is twice the sulfate, giving the beer a sweeter, fuller mouthfeel, especially with that sodium in there as well.

Use a single infusion mash at 69°C (156°F) for 60 minutes or until the starches are adequately converted. This higher mash temperature will leave more dextrins (short-chain polysaccharides that yeast can't metabolize) in the wort, leading to a fuller mouthfeel.

You're shooting for an original gravity of 1.063 (15.5°P), so ensure that your kettle gravity is in the right range based on your evaporation rate.

Boil for 1 hour, adding the hops at the specified times and the fining agent and yeast nutrient with 15 minutes left in the boil.

Add the cacao nibs with 5 minutes left in the boil. Bagging them prior to throwing them in is a good idea, so they don't clog up your system. You can also add them after primary fermentation, but I find adding them at the end of the boil is easy and effective.

Whirlpool for 15 minutes and let stand for an additional 15 minutes.

Chill the beer to fermentation temperature and aerate to 15 ppm oxygen.

Ferment between 18–23°C (64–73°F).

Fermentation

The WLP001 is the OG yeast strain from White Labs. This strain originally derived from the Sierra Nevada Brewery in Chico, California, and was first produced commercially for sale by White Labs in 1995. This is one of the cleanest-fermenting ale yeast strains you will find. Some breweries have even been known to use it for lagers it's so clean. And you can pretty much drive a truck over it and it will still chug through fermentation like a champ. It accentuates hop flavors, which obviously isn't the goal with the beer here, but it also attenuates well, which is good for a big beer like this one. So, despite the high mash temperature, which leaves dextrins behind and a fuller mouthfeel, there will be very little residual sweetness because the yeast will chew through all the available sugars.

We will also typically split this beer into a couple batches and re-ferment one of them with *Brettanomyces* yeast. The *Brettanomyces* has no problem metabolizing those dextrins that the *Saccharomyces* leaves behind, leading to a drier overall beer, with a classic *Brettanomyces* barnyard funk. We sometimes also add some oak that has been soaked in bourbon or even Grand Marnier (that was this year's addition, and it was amazing!). The best part about adding the *Brettanomyces*, though, is the beer then becomes Brett's Chocolate Salty Balls (oh yes we did).

···

KISS MY GRITS SOUTHERN LAGER BY LOST PROVINCE

This is a beer that is regularly produced by our friends at Lost Province Brewing Company in Boone. They were kind enough to share the recipe with us for this book. It's a pre-Prohibition-type lager that is malty and clean, with a little corn flavor to round things out. The recipe is very straightforward, although the corn grits do have to be cooked separately and then added to the mash. It's delicious, and I hope you enjoy it, too!

Ingredients

- 90% 2-Row Pale Ale malt
- 10% corn grits (yellow corn grits from Lindley Mills in Graham, NC)
- Magnum hops—60 minute addition for 9 IBUs
- Cascade hops—whirlpool addition for 5 IBUs
- Fining agent and yeast nutrient optional but recommended
- WLP830 German Lager Yeast

Instructions

The water profile for this beer is very straightforward. Aim for a calcium concentration of at least 50 ppm and a 2:1 chloride to sulfate ratio for a sweeter, fuller mouthfeel that accentuates the maltiness of the beer.

To prepare the grits before adding them to the mash, add enough water to a separate pot to boil the grits. Heat the water until boiling and slowly add in the grits. Stir continuously as they cook for 30 minutes. After 30 minutes, slowly add in cold water to bring the temperature down to 67°C (153°F). As you are doing this, you can prepare your malt for mashing in, so that you can add the grits to the mash after you have mashed in the barley.

Use a single infusion mash at 65°C (149°F) for 60 minutes or until the starches are adequately converted. You will want to start with a thicker mash, with a liquor to grist ratio of 1.6 L water per kg of grain (0.75 quarts of water per pound of grain, as opposed to the typical ratio of 2.6 L water per kg of grain, which translates to 1.25 quarts of water per pound of grain). This adjustment is made to account for the additional water added from the grits. Mix the cooled grits into the mash after the barley has been added. The temperature should stabilize at about 65°C.

You're shooting for an original gravity of 1.045 (11.2°P), so ensure that your kettle gravity is in the right range based on your evaporation rate.

Boil for 1 hour, adding the hops at the specified times and the fining agent and yeast nutrient with 15 minutes left in the boil.

Whirlpool for 15 minutes and let stand for an additional 15 minutes.

Chill the beer to fermentation temperature and aerate to 11 ppm oxygen.

Ferment between 10–13°C (50–55°F).

Fermentation

This is a lager yeast, so the fermentation takes place at a considerably cooler temperature. The German Lager Yeast derives from the Weihen-stephan Brewery in Bavaria, Germany, which is arguably the world's oldest, continuously operating brewery. The exact origin is a bit murky, but it may have been as far back as the twelfth century or even earlier. Regardless, they've been operating for a long time and clearly know what they're doing. Their beers are some of the best examples of classic German styles in my opinion. This particular strain is one of the most widely used lager strains worldwide. It can be used for a number of different types of lagers and produces beautifully clean beers with a slight emphasis on hop character.

PAWPAWPAW

This is a really fun beer that we made recently with our friend Susan Owen, whom we affectionately refer to as the Pawpaw Lady, since she has one of the largest pawpaw orchards in North Carolina, that just so happens to be a short trip down the road from campus. She has beautiful pawpaws and we make beer, which she likes, so it was a perfect match. She cut us a really good deal on some of her pawpaws if we would make her a beer with them. Um, yes please. We thought a Belgian Style Tripel with some fruity esters would be a great way to accentuate the tropical fruit flavors of a pawpaw. And since it's a Tripel, we had to go with the name PawPawPaw. Enjoy!

Ingredients

- 67% 2-Row Pilsner malt
- 10% Light Munich malt
- 10% flaked oats
- 10% sugar (table sugar is fine)
- 3% Carapils/Dextrin malt
- Saaz hops—60 minute addition for 10 IBUs
- Nelson Sauvin hops—whirlpool addition for 20 IBUs
- Pawpaw fruit at 50 g per L to secondary
- Fining agent and yeast nutrient optional but recommended
- WLP530 Abbey Ale Yeast

Instructions

The water profile for this beer is pretty straightforward. Shoot for a calcium concentration of at least 100 pm and a balanced ratio of chloride and sulfate.

Use a single infusion mash at 65°C (149°F) for 60 minutes or until the starches are adequately converted. This mash temperature will leave you with a wort that is highly fermentable, which is what we want. The sugar addition will help to fully attenuate the beer as well. Plus, with such a high original gravity, that sugar addition will help save a little money on malt.

You're shooting for an original gravity of 1.077 (18.7°P), so ensure that your kettle gravity is in the right range based on your evaporation rate.

Boil for 1 hour, adding the hops at the specified times and the fining agent and yeast nutrient with 15 minutes left in the boil.

Whirlpool for 15 minutes and let stand for an additional 15 minutes.

Chill the beer to fermentation temperature and aerate to 19 ppm oxygen.

Ferment between 19–22°C (66–72°F). The lower end of the range will produce an earthier character, whereas the higher end of the range will produce some more fruity ester character.

After primary fermentation is complete, transfer the beer off the yeast to a secondary fermenter and add the pawpaw fruit at the specified quantity. Allow the fruit to ferment for an additional week or until done. Rack off, package, and enjoy.

Fermentation

WLP530 Abbey Ale Yeast originates from Westmalle Trappist Brewery in Belgium. There are only a handful of Trappist breweries in the world (somewhere around 13 as of the writing of this book). These are Cistercian monasteries in which the monks actually brew the beer at the monastery. Westmalle was one of the OG Trappist breweries and is even credited with coining the name Tripel for the golden strong pale ale they made. And, in my opinion, their Tripel is the one that everyone should be trying to em-ulate. The fruity esters produced by this strain are reminiscent of cherry, plum, and pear. I like this strain because it produces less of the bubblegum character that some of the other Belgian yeast strains produce. This strain is relatively flocculant as well, so you should be left with a fairly clear beer.

MALT VINEGAR

Now that you've made all this beer and have a bunch left over that you don't know what to do with, let's turn some of it into malt vinegar!

Ingredients

- Leftover beer (beer with an ABV in the range of 5–7% is ideal)
- Live culture vinegar added at 10–20% of the mass of the beer

Instructions

Add leftover beer to a large glass jar or similar container.

Add the live culture vinegar at the specified quantity to the beer.

You can cover this with a towel secured with a rubber band and allow it to ferment at room temperature for several months. However, a much more efficient way to make vinegar is to use an aquarium pump with an air stone on the end for diffusing bubbles into the solution. Place the air stone in the jar with the inoculated beer. Turn on the pump, cover the jar with a towel secured with a rubber band, and leave it at room temperature to ferment. This reduces the overall time to make vinegar from a few months to 1–2 weeks.

After 1 week (if using the pump), check the pH of the vinegar and taste it.

Once the vinegar is in the 3.5–4.0 pH range, the vinegar should be done. However, taste is the best guide for knowing when it is ready. If it tastes good, it's ready.

Remove the pump, cover the jar with a lid, and store the vinegar in the refrigerator.

Fermentation

The acetic acid bacteria from the live culture vinegar convert the ethanol in the beer to acetic acid, the acid that makes vinegar. Acetic acid bacteria are aerobic, so they need access to oxygen. This is why you cover the jar with a towel, so that oxygen can still permeate the headspace of the jar. And it's also why pumping in air is so effective at reducing the time of this fermentation. Most vinegars are about 5% acetic acid, so that's why you want to shoot for a similar alcohol range. Most of the alcohol will be converted to acetic acid. This is also why it's important to check the pH and taste the vinegar. You don't want it to get overly acidic. You can also do this with any alcoholic beverage, so don't limit yourself to beer. Just make sure you don't start with anything overly alcoholic, though, or you'll end up with an acetic acid bomb. Enjoy!

Conclusion

HEALTHFULNESS OF MICROBES AND CONCLUDING REMARKS

———

We are only beginning to understand the incredible complexity of the microbial world and our relationship with the microbes in our environment. The more we learn, however, the more we discover how important a healthy gut microbiome is to our overall physical and mental health. It has been evident for quite some time that there is a relationship between mental and gastrointestinal health, as is evident, for example, in anxiety-induced gastric distress or in stress eating. More recent studies have begun to establish the neural, endocrine, and immune pathways that are used for communication between the gut and the central nervous system (Cryan and Dinan 2012). But beyond the correlations, there are two main questions that must be answered to better understand these relationships. First, are there particular changes in the gut microbiome that can be associated with these gastric issues? And if so, is it the deterioration in mental health that causes the change in the gut microbiome, or do changes in the gut microbiome lead to a deterioration in mental health? Surprisingly, the second question has thus far been the easier nut to crack. And it appears that it's definitely the case that changes in the gut microbiome appear to induce changes in mental health. When fecal microbiota were transplanted from clinically depressed human patients to rats that had their own gut microbiota depleted prior to transplantation, the rats began to display behaviors characteristic of depression (Kelly et al. 2016). However, it also seems to go both ways. For example, stress or just being overworked can lead to poor eating habits, which in turn can cause changes to the gut microbiota that may then exacerbate mental health deterioration.

More recent studies have begun to answer the first question as well. A study published in *Nature Communications* in 2022 using a large European cohort found 12 genera and one family of bacteria that were associated with symptoms of depression (Radjabzadeh et al. 2022). This was not a big surprise, since the microbes identified by the study are involved in the production of critical neurotransmitters for depression, namely serotonin, butyrate, glutamate, and gamma amino butyric acid (GABA). These small signaling molecules as well as certain short-chain fatty acids have become the focus of research into many neurological conditions, since their proper concentrations in the body seem to determine proper neurological function. Too little or too much and things start to go haywire. And your gut microbes are largely responsible for producing them. In individuals with depressive symptoms, all the microbes identified in the study were depleted other than the genera *Sellimonas, Eggerthella, Lachnoclostridium*, and *Hungatella*, which were found in greater abundance in individuals with depressive symptoms. *Eggerthella* is of particular interest as not only have several other studies found it to be increased in people with depressive symptoms, but the authors of the most recent study also used a statistical analysis to identify a causal link with symptoms of depression in the cohort studied.

We still have a long way to go before we have a more thorough understanding of these very complex interrelationships, but it is likely that we will begin to see more directed probiotic treatments for mental health care and neurological issues in the near future. In the meantime, like your mother used to tell you, "Eat your vegetables." A healthy, balanced diet that is high in fiber, with plenty of whole grains and vegetables, and with limited animal protein (Mediterranean diet anyone?), supports the healthy gut bacteria that are key to producing the proper signaling molecules and promoting healthy neurological function. The high-fiber component may be the most critical element in supporting the growth of the proper gut bacteria, as the fiber acts as a prebiotic that the correct microbes consume. And you'll be more regular and potentially lower your cholesterol at the same time, so why not.

Aside from mental health, our gut microbiome also plays a critical role in our overall physical health. Roughly 70% of the immune system is located in the gut, and the immune cells rely on healthy interactions with gut bacteria for proper function (Callahan 2023). Obesity and gastrointestinal cancers are both characterized by certain molecular mechanisms that lead to inflammation and ultimately disease. The composition of gut bacteria and their resultant metabolic activity has been associated with gastrointestinal cancer. Additionally, the gut bacteria that produce short-chain fatty acids or support the gut barrier function appear to provide a therapeutic benefit to reduce the chronic inflammation that may lead to disease (Cani and Jordan 2018).

Not to fear, though. A well-balanced, high-fiber diet supports the proper gut microbes to maintain healthy immune function as well. Also, a diet high in fermented foods increases gut microbial diversity and reduces inflammatory markers (Wastyk et al. 2021). So, go out there and buy some yogurt and kimchi and kombucha. These are all great sources of probiotics (as long as the products you buy have live cultures—check the labels!) that will support your gut microbial diversity. Even better, though, is to ferment your own products at home. By doing so, you are incorporating some of the microbial diversity from your immediate environment (especially if you're using homegrown or locally grown fruits and vegetables for fermenting) into your gut rather than relying on microbes from wherever the commercially obtained fermented products were prepared.

Our immediate environments appear to have a large impact on our gut microbial makeup. We are born as essentially microbial blank slates until we are "seeded" with microbes by our mothers. The microbial diversity we receive after birth lasts throughout our lives, but it diminishes over time as we interact less closely with our mothers and eventually move out (hopefully). A recent study published in *Nature* showed that our immediate environments and who we cohabitate with (and for how long) then have large influences on our gut microbiome (Valles-Colomer, Blanco-Míguez and Manghi 2023). There was, however, no difference observed in transmission among individuals in Western and non-Western cultures, despite apparent differences in hygiene and living conditions between the two populations. This fact shows that the greater diversity of gut microbes in individuals in non-Western cultures is a result of diet and interactions with their environment rather than dependent upon transmission among individuals. Encourage the healthful bacteria in your environment by growing them in fermented products you make at home and incorporating them into your gut microbiome. In short, make friends with the bacteria in your environment and share them among your friends and cohabitants. Everyone will thank you (at least they won't curse you for making them sick).

When I initially sat down to write this book, I imagined the symmetry of beginning with the difficulty of defining what encompasses fermentation and ending with my witty take on what it means. At the end of the day, though, who cares what definition of fermentation you prefer (which is my way of admitting that I haven't settled on such a witty definition). More important is that you understand the basics of the process so that you can practice it safely and effectively, which I sincerely hope this book has helped with. Make peace, not war, with the microbes in your environment, and learn to nurture the ones covered in this book, so that you can work with them to unlock the amazing spectrum of flavors and nutritional attributes that only fermentation can supply. Now, time to put the book down and get fermenting. Here's to a healthier, fermented future. As we like to say in our program, research and education never tasted so good!

Appendix

UNIT CONVERSIONS

In case you hadn't already figured this out, I'm not a fan of the customary units we use in this country. The metric system, based on units of 10, is so much easier for doing calculations. Not to mention the fact that literally the rest of the world uses the metric system (which, I suspect, is the real reason why we [as Americans] don't use it). I guess we just like to prove our affinity for difficult calculations in this country. The Fahrenheit temperature scale is equally nonsensical. Why is the freezing point of water 32° and the boiling point 212°? The short answer is because of the scale that the German physicist Daniel Gabriel Fahrenheit used to set his measured temperatures. Which figures, considering physicists are notoriously loosey-goosey with their units. Again, though, for some reason we have chosen to use the illogical scale that nobody else in the world uses. The Celsius scale is so much more logical. The freezing point of water is 0° and boiling is 100°. Now, there's a scale I can wrap my brain around. Plus, it's the same scale as the absolute temperature scale, Kelvin, just offset by 273 degrees, which is why we use it in the scientific world.

TABLE A.1. Temperature conversions

Unit	Conversion formula
°F	(1.8 × °C) + 32
°C	(°F − 32) × 5/9

Also, we need to stop measuring solids in volumes. The mass of a teaspoon of salt will vary greatly depending on the size of the salt crystals, rendering the volumetric measurement meaningless. In general, gravimetric (by mass) measurements are far more accurate than volumetric ones. So, you should even be measuring your liquids by mass rather than by volume. Remember, the density of water is 1 gram per milliliter (this is temperature dependent, but it's close enough at room temperature to use this for converting volumes of water to masses).

TABLE A.2. Customary unit to metric unit conversions

Customary unit	Metric unit	Approx. value	Mass
1 teaspoon	4.93 mL	5 mL	4.93 g
1 tablespoon (3 teaspoons)	14.79 mL	15 mL	14.79 g
1 cup (8 fluid ounces)	240 mL	—	240 g
1 pint (2 cups)	480 mL	—	480 g
1 quart (2 pints, 4 cups)	960 mL	1 L	960 g
1 gallon (4 quarts)	3.785 L (3,785 mL)	4 L	3.785 kg
1 fluid ounce	29.6 mL	30 mL	30 g
1 ounce (mass)	28.35 g	28 g	—
1 pound (16 ounces)	453.6 g	454 g	—

WORKS CITED

Alba-Lois, L., and C. Segal-Kischinevzky. 2010. "Beer and Wine Makers." *Nature Education*, 17.

Bai, B., F. Chen, Z. Wang, X. Liao, and G. Zhao. 2005. "Mechanism of the Greening Color Formation of Laba Garlic, a Traditional Homemade Chinese Food Product." *Journal of Agricultural and Food Chemistry* 53 (18): 7103–7.

Bai, B., L. Li, X. Hu, Z. Wang, and G. Zhao. 2006. "Increase in the Permeability of Tonoplast of Garlic (Allium sativum) by Monocarboxylic Acids." *Journal of Agricultural and Food Chemistry* 54 (21): 8103–7.

Baron, Stanley. 1960. *Brewed in America A History of Beer and Ale in the United States*. Boston: Little, Brown.

Barthole, Jenn. 2022. "Black-Owned, Award Winning Whiskey Brand 'Uncle Nearest' Tops Esteemed Inc. 5000 List." *Ebony*, August 25, 2022. www.ebony.com/black -owned-award-winning-whiskey-brand-uncle-nearest-lands-on-esteemed-inc -5000-list.

Berger, Arthur Asa. 1970. *Li'l Abner: A Study in American Satire*. Jackson: University Press of Mississippi.

Bryant, K. L., C. Hansen, and E. E. Hecht. 2023. "Fermentation Technology as a Driver of Human Brain Expansion." *Communications Biology* 6 (November): 1190. https://doi.org/10.1038/s42003-023-05517-3.

Callahan, Alice. 2023. "The Wild World Inside Your Gut." *New York Times*, February 23, 2023.

Cani, Patrice D., and Benedicte F. Jordan. 2018. "Gut Microbia-Mediated Inflammation in Obesity: A Link with Gastrointestinal Cancer." *Nature Reviews Gastroenterology and Hepatology* 15 (11): 671–82.

Collins, Lewis, and Richard Collins. 1874. *History of Kentucky, Vol. 1*. Salem, MA: Higginson Book Company.

Cryan, John F., and Timothy G. Dinan. 2012. "Mind-Altering Microorganisms: The Impact of the Gut Microbiota on Brain and Behaviour." *Nature Reviews Neuroscience* 13 (10): 701–12.

Darby, T. M., J. A. Owens, B. J. Saeedi, L. Luo, J. D. Matthews, B. S. Robinson, C. R., Jones, and R. M. Naudin. 2019. "*Lactococcus Lactis* Subsp. *cremoris* Is an Efficacious Beneficial Bacterium That Limits Tissue Injury in the Intestine." *iScience*, 356–67. https://doi.org/10.1016/j.isci.2019.01.030.

Dietsch, Michael. 2016. *Whiskey: A Spirited Story with 75 Classic and Original Cocktails*. New York: Countryman Press.

Diez-Ozaeta, Iñaki, and Oihana Juaristi Astiazaran. 2022. "Recent Advances in Kombucha Tea: Microbial Consortium, Chemical Parameters, Health Implications and Biocellulose Production." *International Journal of Food Microbiology* 377 (September): 109783.

DiNicolantonio, James J., and James H. O'Keefe. 2017. "Good Fats versus Bad Fats: A Comparison of Fatty Acids in the Promotion of Insulin Resistance, Inflammation, and Obesity." *Missouri Medicine* 114, no. 4 (July–August): 303–7.

Donnelly, Catherine W. 2014. *Cheese and Microbes*. Washington, DC: American Society for Microbiology.

Edwards, Megan E. 2011. "Virginia Ham: The Local and Global of Colonial Foodways." *Food and Foodways* 19 (1–2): 56–73.

Egbuta, M. A., M. Mwanza, and O. O. Babalola. 2017. "Health Risks Associated with Exposure to Filamentous Fungi." *International Journal of Environmental Research and Public Health* 14, no. 7 (July): 719.

Egerton, John. 1993. *Southern Food: At Home, on the Road, in History*. Chapel Hill: University of North Carolina Press.

Fishwick, Marshall. 1964. "Southern Cooking." In *American Heritage Cookbook and Illustrated History of American Eating and Drinking*, by Cleveland Armory, Lucius Beebee, Evan Jones, Archie Robertson, Leonard Louis Levinson, Russell Lynes, and Paul Engle; edited by Helen Duprey Bullock and Helen McCully, 629. New York: Simon and Shuster.

Fluker, Dominique. 2022. "How Fawn Weaver Created Uncle Nearest Premium Whiskey from Hidden History." *Forbes*, March 9, 2022.

González-Quijano, G. K., L. Dorantes-Álvarez, H. Hernández-Sánchez, M. E. Jaramillo-Flores, M. de Jesús Perea-Flores, A. Vera-Ponce de León, and C. Hernández-Rodríguez. 2014. "Halotolerance and Survival Kinetics of Lactic Acid Bacteria Isolated from Jalapeño Pepper (*Capsicum annuum* L.) Fermentation." *Journal of Food Science* 79 (8): M1545–53.

Green, Ben A. 2017. *Jack Daniel's Legacy*. 50th anniv. ed. Nashville, TN: Grant Sidney Publishing.

Hadi, Amir, Makan Pourmasoumi, Ameneh Najafgholizadeh, Cain C. T. Clark, and Ahmad Esmaillzadeh. 2021. "The Effect of Apple Cider Vinegar on Lipid Profiles and Glycemic Parameters: A Systematic Review and Meta-analysis of Randomized Clinical Trials." *BMC Complementary Medicine and Therapies* 21 (1): 179. https://doi.org/10.1186/s12906-021-03351-w.

Halake, Niguse, and Bhaskar Chinthapalli. 2020. "Fermentation of Traditional African Cassava Based Foods: Microorganisms Role in Nutritional and Safety Value." *Journal of Experimental Agriculture International* 42 (9): 56–65.

Horowitz, B. Z. 2005. "Botulinum Toxin." *Critical Care Clinics* 21 (4): 825–39.

H. S. Rich and Co. 1974. *One Hundred Years of Brewing: A Complete History of Progress Made in Art Science and Industry of Brewing in the World*. New York: Arno Press.

Humpenöder, F., B. L. Bodirsky, I. Weindl, H. Lotze-Campen, T. Linder, and A. Popp. 2022. "Projected Environmental Benefits of Replacing Beef with Microbial Protein." *Nature* 605:90–96.

Imai, S., K. Akita, M. Tomotake, and H. Sawada. 2006. "Identification of Two Novel Pigment Precursors and a Reddish-Purple Pigment Involved in the Blue-Green Discoloration of Onion and Garlic." *Journal of Agricultural and Food Chemistry* 54 (3): 843–47.

Jeong, J. K., Y. W. Kim, H. S. Choi, D. S. Lee, S. A. Kang, and K. Y. Park. 2011. "Increased Quality and Functionality of Kimchi When Fermented in Korean Earthenware (Onggi)." *International Journal of Food Science and Technology* 46 (10): 2015–21.

Jones House. n.d. "Boone Creek (Kraut Creek): Historic Downtown Boone Walking Tour." PocketSights. Accessed 2023. https://pocketsights.com/tours/place/Boone-Creek-%28Kraut-Creek%29-28007:3449.

Karwowska, M., and A. Kononiuk. 2020. "Nitrates/Nitrites in Food-Risk for Nitrosative Stress and Benefits." *Antioxidants (Basel)* 9 (3): 241–58.

Keene, Sarah, Manbeer S. Sarao, Philip J. McDonald, and Jennifer Veltman. 2019. "Cutaneous Geotrichosis Due to *Geotrichum candidum* in a Burn Patient." *Access Microbiology* 1, no. 1 (March): e000001. https://doi.org/10.1099/acmi.0.000001.

Kelly, John R., Yuliya Borre, Ciaran O'Brien, Elaine Patterson, Sahar El Aidy, Jennifer Deane, Paul J. Kennedy, et al. 2016. "Transferring the Blues: Depression-Associated Gut Microbiota Induces Neurobehavioural Changes in the Rat." *Journal of Psychiatric Research* 82 (November): 109–18.

Kennell, Tiana. 2023. "Highland Brewing's Oscar Wong Awarded Top NC Civilian Honor, Order of the Long Leaf Pine." *Asheville Citizen Times*, May 5, 2023.

Kim, Soohwan, and David L. Hu. 2023. "Onggi's Permeability to Carbon Dioxide Accelerates Kimchi Fermentation." *Journal of the Royal Society Interface* 20, no. 201 (April): 20230034. https://doi.org/10.1098/rsif.2023.0034.

Kindstedt, Paul S. 2013. "The Basics of Cheesemaking." In *Cheese and Microbes*, edited by Catherine W. Donnelly, 17–38. Washington, DC: American Society of Microbiology.

Kitamoto, Katsuhiko. 2015. "Cell Biology of the Koji Mold *Aspergillus oryzae*." *Bioscience, Biotechnology, and Biochemistry* 79, no. 6 (June): 863–69.

Kitwetcharoen, Haruthairat, Ly Tu Phung, Preekamol Klanrit, Sudarat Thanonkeo, Patcharaporn Tippayawat, Mamoru Yamada, and Pornthap Thanonkeo. 2023. "Kombucha Healthy Drink—Recent Advances in Production, Chemical Composition and Health Benefits." In "Research Advances in Fermented Beverages," edited by Zhao Jin. Special issue, *Fermentation* 9, no. 1 (January): 48. ttps://doi.org/10.3390/fermentation9010048.

Kurlansky, Mark. 2002. *Salt: A World History*. New York: Walker and Co.

Lane, Nick. 2015. "The Unseen World: Reflections on Leeuwenhoek (1677) 'Concerning Little Animals.'" *Philosophical Transactions of the Royal Society B: Biological Sciences* 370, no. 1666 (April): 0140344.

Lawson, John. 1967. *A New Voyage to Carolina*. Edited by Hugh Talmage Lefler. Chapel Hill: University of North Carolina Press.

Lee, Soomin, Heeyoung Lee, Sejeong Kim, Jeeyeon Lee, Jimyeong Ha, Yukyung Choi, Hyemin Oh, Kyoung-Hee Choi, and Yohan Yoon. 2018. "Microbiological Safety of Processed Meat Products Formulated with Low Nitrite Concentration—A Review." *Asian-Australasian Journal of Animal Sciences* 31 (8): 1073–77.

Leigh, Meredith. 2018. *Pure Charcuterie: The Craft and Poetry of Curing Meat at Home*. Gabriola Island, BC: New Society Publishers.

———. 2020. *The Ethical Meat Handbook: From Sourcing to Butchery, Mindful Meat Eating for the Modern Omnivore*. Gabriola Island, BC: New Society Publishers.

Leistner, L. 1992. "The Essentials of Producing Stable and Safe Raw Fermented Sausages." In *New Technologies for Meat and Meat Products: Fermentation and Starter Cultures, Muscle Enzymology and Meat Ageing, Quality Control Systems*, edited by Frans J. M. Smulders, 1–19. Utrecht, Netherlands: ECCEAMEST.

Levinovitz, Alan. 2015. *The Gluten Lie: And Other Myths about What You Eat*. New York: Regan Arts.

Libkind Diego, Chris Todd Hittinger, Elisabete Valério, Carla Gonçalves, Jim Dover, Mark Johnston, Paula Gonçalves, and José Paul Sampaio. 2011. "Microbe Domestication and the Identification of the Wild Genetic Stock of Lager-Brewing Yeast." *Proceedings of the National Academy of Sciences* 108, no. 35 (August): 14539–44.

Ligenza, Alicja, Karolina Patrycja Jakubczyk, Joanna Kochman, and Katarzyna Janda. 2021. "Health-Promoting Potential and Microbial Composition of Fermented Drink Tepache." *General Medicine and Health Sciences* 27 (3): 272–76.

Li Liu, Jiajing Wang, Danny Rosenberg, Hao Zhao, György Lengyel, and Dani Nadel. 2018. "Fermented Beverage and Food Storage in 13,000 Y-old Stone Mortars at Raqefet Cave, Israel: Investigating Natufian Ritual Feasting." *Journal of Archaeological Science: Reports* 21 (October): 783–93.

Mariani, John F. 2014. *The Encyclopedia of American Food and Drink*. London: Bloomsbury Publishing.

Matsuura, Noriko, Hidemasa Motoshima, Kenji Uchida, and Yujiro Yamanaka. 2022. "Effects of *Lactococcus lactis* Subsp. *cremoris* YRC3780 Daily Intake on the HPA Axis Response to Acute Psychological Stress in Healthy Japanese Men." *European Journal of Clinical Nutrition* 76 (4): 574–80.

McDaniel, Rick. 2015. "Recipe for Success: For Decades, Boone Was the Capitol of North Carolina–Made Sauerkraut." *WNC Magazine*, November 2015. https://wncmagazine.com/feature/recipe_success.

Mintz, Sidney W. 1986. *Sweetness and Power: The Place of Sugar in Modern History*. New York: Penguin Books.

Mitenbuler, Reid. 2015. *Bourbon Empire: The Past and Future of America's Whiskey*. New York: Penguin Random House.

Müller, Alexandra, Niels Rösch, Gyu-Sung Cho, Ann-Katrin Meinhardt, Jan Kabisch, Diana Habermann, Christina Böhnlein, Erik Brinks, Ralf Greiner, and Charles M. Franz. 2018. "Influence of Iodized Table Salt on Fermentation Characteristics and Bacterial Diversity during Sauerkraut Fermentation." *Food Microbiology*, no. 76 (December): 473–80.

Nurkolis, Fahrul, Faqrizal Ria Qhabibi, Vincentius Mario Yusuf, Stanely Bulain, Ghevira Naila Praditya, Deogifta Graciani Lailossa, Msy Firyal Nadya Al Mahira, et al. 2022. "Anticancer Properties of Soy-Based Tempe: A Proposed Opinion for Future Meal." *Frontiers in Oncology* 12 (October). https://doi.org/10.3389/fonc.2022.1054399.

Orton, V. 1995. *The American Cider Book: The Story of America's National Beverage.* New York: North Point Press.

Pharms, Gabrielle Nicole. 2022. "Drinks Innovators of the Year: Fawn Weaver and Victoria Eady Butler." *Food and Wine,* March 15, 2022.

Radjabzadeh, Djawad, Jos A. Bosch, André G. Uitterlinden, Aeilko H. Zwinderman, M. Arfan Ikram, Joyce B. J. van Meurs, Annemarie I. Luik, et al. 2022. "Gut Microbiome-Wide Association Study of Depressive Symptoms." *Nature Communications* 13 (December): 1728. https://doi.org/10.1038/s41467-022 -34502-3.

Regan, Gary, and Mardee Haidin Regan. 2009. *The Book of Bourbon and Other Fine American Whiskeys.* New York: Jared Brown.

Sagan, Lynn. 1967. "On the Origin of Mitosing Cells." *Journal of Theoretical Biology* 14, no. 3 (March): 225–74.

Schoene, Lorin, Ellis I. Fulmer, and L. A. Underkofler. 1940. "Saccharification of Starchy Grain Mashes for the Alcoholic Fermentation Industry." *Industrial and Engineering Chemistry* 32 (4): 544–47.

Smith, Gregg. 1998. *Beer in America: The Early Years—1587–1840.* Boulder, CO: Brewers Publications.

Stephenson, Frank Jr., and Barbara Nichols Mulder. 2017. *North Carolina Moonshine: An Illicit History.* Charleston, SC: American Palate.

Takamine, Jokichi. 1914. "Enzymes of *Aspergillus oryzae* and the Application of Its Amyloclastic Enzyme to the Fermentation Industry." *Journal of Industrial and Engineering Chemistry* 6 (10): 824–28.

Tamam, Badrut, Dahrul Syah, Maggy Thenawidjaja Suhartono, Wisnu Ananta Kusuma, Shinjiro Tachibana, and Hanifah Nuryani Lioe. 2019. "Proteomic Study of Bioactive Peptides from Tempe." *Journal of Bioscience and Bioengineering* 128, no. 2 (August): 241–48.

Teniola, O. D., and S. A. Odunfa. 2002. "Microbial Assessment and Quality Evaluation of Ogi during Spoilage." *World Journal of Microbiology and Biotechnology* 18:731–37.

Toldra, Fidel. 2012. "Biochemistry of Fermented Meat." In *Food Biochemistry and Food Processing,* edited by Benjamin K. Simpson, 331–43 Ames, IA: Wiley-Blackwell.

Toldra, Fidel, Y. H. Hui, Iciar Astiasaran, Joseph G. Sebranek, and Régine Talon. 2015. *Handbook of Fermented Meat and Poultry.* West Sussex: John Wiley and Sons.

USDA (US Department of Agriculture). 1923. *United States Department of Agriculture Yearbook 1922.* Washington, DC: Government Printing Office.

Valles-Colomer, Mireia, Aitor Blanco-Míguez, Paolo Manghi, Francesco Asnicar, Leonard Dubois, Davide Golzato, Federica Armanini, et al. 2023. "The Person-to-Person Transmission Landscape of the Gut and Oral Microbiomes." *Nature* 614, no. 7946 (January): 125–35.

Veach, Michael R. 2013. *Kentucky Bourbon Whiskey.* Lexington: University Press of Kentucky.

Wall, Tamara L., and Cindy L. Ehlers. 1995. "Genetic Influences Affecting Alcohol Use among Asians." *Alcohol Health and Research World* 19 (3): 184–89.

Wastyk, Hannah C., Gabriela K. Fragiadakis, Dalia Perelman, Dylan Dahan, Bryan D. Merrill, B. Yu Feiqiao, Madeline Topf, et al. 2021. "Gut-Microbiota-Targeted Diets Modulate Human Immune Status." *Cell* 184 (16): 4137–53.

Watson, B. 1999. *Cider, Hard and Sweet: History, Traditions, and Making Your Own.* Woodstock, VT: Countryman Press.

Watts, E. G., M. E. Janes, W. Prinyawiwatkul, Y. Shen, Z. Xu, and D. Johnson. 2018. "Microbiological Changes and Their Impact on Quality Characteristics of Red Hot Chili Pepper Mash during Natural Fermentation." *International Journal of Food Science and Technology* 53 (8): 1816–23.

Williams, T., D. Parker, and B. Taubman. 2021. "Characterization of Unmalted Barley Treated with *Aspergillus oryzae*." *Journal of the American Society of Brewing Chemists* 80 (4): 427–34.

Wolfe, Benjamin. 2023. "Geotrichum Candidum: A Yeast Holding on to Its Moldy Past." Microbialfoods.org, December 15, 2023. https://microbialfoods.org /geotrichum-candidum-mold-transition.

Zeuthen, P. 2015. "A Historical Perspective of Meat Fermentation." In *Handbook of Fermented Meat and Poultry*, edited by Fidel Toldra, Y. H. Hui, Iciar Astiasarán, Joseph G. Sebranek, and Régine Talon, 1–8. Ames, IA: Blackwell Publishing.

INDEX

Page numbers in italics refer to illustrations.

ABOUT THE AUTHOR

Dr. Brett Taubman is a professor in the A.R. Smith Department of Chemistry and Fermentation Sciences at Appalachian State University. He earned his BS degrees in finance and chemistry from the Pennsylvania State University and Montana State University, respectively, and his PhD in analytical chemistry from the University of Maryland in 2004. Following his graduate studies, he worked as a postdoctoral research associate at the Pennsylvania State University before joining the faculty at Appalachian State University in 2007. Dr. Taubman has found his research niche as a fermentation chemist, exploring the fascinating worlds of beer, food, and mold chemistry. He has successfully developed an instructional fermentation facility on campus and serves as President of Ivory Tower, Inc., a 501(c)(3) nonprofit corporation with the mission of supporting research and education within fermentation sciences. Dr. Taubman helped to develop the four-year degree program in Fermentation Sciences, for which he is currently the director. When not educating the future innovators of the American fermentation industries, he enjoys keeping his lovely wife and furry friends happy, hiking with them in the beautiful southern Appalachian Mountains (with the ulterior motive of foraging for fungi); pretending that he's still young by playing soccer with the lads and shredding the gnar; and trying occasionally to see his two adult sons who have inexplicably settled in the Midwest.